Springer Texts in Business and Economics

Springer Texts in Business and Economics (STBE) delivers high-quality instructional content for undergraduates and graduates in all areas of Business/Management Science and Economics. The series is comprised of self-contained books with a broad and comprehensive coverage that are suitable for class as well as for individual self-study. All texts are authored by established experts in their fields and offer a solid methodological background, often accompanied by problems and exercises.

Dietmar Ernst • Joachim Häcker

Corporate Risk Management

A Case Study on Risk Evaluation

Springer

Dietmar Ernst
Nürtingen-Geislingen University
of Applied Science (HfWU)
Nürtingen, Germany

Joachim Häcker
Munich University of Applied Sciences
Munich, Germany

ISSN 2192-4333 ISSN 2192-4341 (electronic)
Springer Texts in Business and Economics
ISBN 978-3-031-53125-5 ISBN 978-3-031-53126-2 (eBook)
https://doi.org/10.1007/978-3-031-53126-2

Translation from the German language edition: "Risikomanagement im Unternehmen Schritt für Schritt" by Dietmar Ernst and Joachim Häcker, © UTB GmbH 2021. Published by UTB GmbH. All Rights Reserved.

This Springer imprint is published by the registered company Springer Nature Switzerland AG
The registered company address is: Gewerbestrasse 11, 6330 Cham, Switzerland

Preface

Risk management is one of the major topics in business and academic education. Pandemics, economic crises, disruptions caused by new technologies, and competitive risks require entrepreneurial decisions to set up companies for long-term success. These decisions must be well informed. Since future developments cannot be predicted with certainty, the risks (in the sense of opportunities and threats) to the expected earnings must be taken into account in the decision-making process.

Risk management supports a company's management in recognizing and dealing with significant risks that could endanger the company's success or future existence. Risk management includes the following tasks for evaluating Risk:

- Risk identification
- Risk quantification
- Risk aggregation
- Risk analysis
- Risk assessment
- Risk hedging

Risk management is a very complex subject matter. It is characterized by:

- Numerous legal and quasi-legal requirements
- Quantitative know-how, which requires statistical knowledge and an understanding of risk models, and
- The expertise in IT tools to implement the above-mentioned risk management tasks.

In a study conducted by Deloitte in 2020, 64 risk managers of German companies were interviewed. The participants in the study were asked about the risk management systems used in their companies in relation to the amendments to IDW PS 340. The results were as follows:

Risk-bearing capacity concepts of the companies surveyed: More than half of the risk managers surveyed stated that their companies do not currently have a concept for deriving risk-bearing capacity.

Risk aggregation: 43% of the participants stated that they determine their overall risk position by adding up the loss expectations of individual risks. The fact that the addition of individual risks is not a suitable measure for determining the overall risk of a corporation either was not known or ignored. Only 24% of the companies surveyed were using a suitable quantitative aggregation method in the form of a simulation procedure.

In academic training and continuing professional development, it has been observed for years that the interest and competence in quantitative topics have been declining sharply. Even in the training of risk managers, the scope of quantitative methods has been reduced, although the quantification and aggregation of risks precisely require these quantitative competencies.

The reason for the declining willingness to deal with quantitative methods in risk management is not due to the risk managers or students themselves, but rather due to the way in which quantitative training is carried out.

When risk managers and students are taught quantitative risk management from a mathematical or statistical perspective, they are confronted with materials that are usually overloaded with formulas that cannot be easily understood, even by an economist. Furthermore, the reference to practical scenarios is often missing. On the other hand, teachings from an economic perspective only deal with quantitative methods at a broad level, where more emphasis is placed on qualitative risk management. There is a gap between the coverage of the two, which makes it difficult for risk managers and students to acquire deep understanding of quantitative risk management.

This textbook and practitioner's guide aims to close this gap. It forms the basis for the training to become a "Certified Financial Engineer (CFE)" with a specialization in "Risk Management", and our MBA in "Applied Quantitative Finance". The following philosophy is applied:

- Audit standard compliant: The present textbook and the "Certified Financial Engineer (CFE)" teaches all competences required by audit standards, from risk analysis to risk aggregation with Monte Carlo simulation.
- Case study-based training: This textbook and course uses one case study throughout that depicts tasks in practical risk management. This means that the key learnings from this book can be immediately applied in professional practice.
- High learning success: The case study is divided into small tasks that can be worked on in quick succession. Hence, the learning success is immediately visible.
- Level-oriented advanced training: Each participant can take the certificate course at the level at which they would like to advance their training. Both beginners and experienced risk managers will find this as a very good return on their investment.
- Flexible learning: The CFE is a digital course. You do not have to think about physical attendance and can work on the case studies anytime and anyplace.

- Modelling-based risk management: The approach in this course is model-based risk management. This means everything can be implemented directly in Excel including the use of Excel add-ins. This allows the course to be accessible to anyone as users do not need to buy specialized software to become a CFE and is easily applicable in the workplace.
- Minimal prior knowledge required: You only need minimal knowledge of mathematics and statistics before starting this course. The content is explained step-by-step. Experienced risk managers can build on this base as required.
- Personal support: Our philosophy is that advanced training is only successful if it is fun and motivating. Due to the didactic preparation of the course and the useful tips given throughout, learning success is guaranteed.
- Feedback through checkpoints: You will pass checkpoints during the course that provide direct feedback on the models you have created yourself.
- Professional course materials: You will receive high-quality course materials, Excel sample models and detailed descriptions of exact solutions.
- Accredited programme: The contents of the "Certified Financial Engineer (CFE)" are accredited and are also used in the MBA module of "Applied Quantitative Finance".
- German or English: You can complete the programme in either German or English.
- Recognized title: The title "Certified Financial Engineer (CFE)" and MBA in "Applied Quantitative Finance" is highly reputable among companies and opens up interesting career opportunities.

In teaching and professional development, it has been exhibited that our modelling and case study-based approach to risk management helps all students, professionals and any other participants, to understand the topic of quantitative risk management easily. We are always pleased to see the enthusiasm and motivation with which our participants and students work on the case studies and witness their advancing practical knowledge. Since quantitative risk management provides a strong base and has a strong correlation to professional functions of controlling, corporate finance, derivatives, and portfolio management, it opens up a variety of very good career opportunities. In a world of increasing digitalization and data analytics, risk managers and students with a quantitative education background are in demand. You can find more information about our programmes by visiting the following websites:

www.eiqf.de
www.certified-financial-engineer.de

We would also like to take this opportunity to thank all those who provided us with technical support during the preparation of the book. Our special thanks go to Prof. Dr. Frank Altrock, Mr. Alexander Stewart and Mr. Shwetabh Singh. You can reach us at program@eiqf.de.

Nürtingen, Germany Dietmar Ernst
Munich, Germany Joachim Häcker

Contents

General Structure of the Case Study

The case study is divided into two courses. Each course is divided into three course units.

Course 1 deals with risk analysis and is divided into the following course units:

- Course unit 1: Graphical illustration of risks
- Course unit 2: Variance and standard deviation
- Course unit 3: Models for calculating volatility

Course 2 contains the application of quantitative instruments in risk management and risk analysis. It is divided into the following course units:

- Course unit 1: Different types of Value at Risk and Lower Partial Moments and Extreme Value Theory
- Course unit 2: Determination of portfolio risks
- Course unit 3: Hedging of hedgeable risks and modelling of non-hedgeable risks

Detailed Structure of the Case Study

Course 1: Risk analysis
Course unit 1: Graphical illustration of risks

- You receive a data set consisting of three assets representing commodity price change risk, exchange rate risk, and interest rate risk.
- You will get to know the different calculations of discrete and continuous returns and will be able to calculate them for different periods of time.

(continued)

D. Ernst, J. Häcker, *Corporate Risk Management*, Springer Texts in Business and Economics, https://doi.org/10.1007/978-3-031-53126-2_1

Course 1: Risk analysis (continued)

- You will be able to represent continuous and discrete returns graphically and explain the statistical concepts behind them.
- You will be able to calculate skewness and kurtosis and interpret their contents.

Course unit 2: Variance and standard deviation

- You will be calculating the variance and standard deviation as well as the semivariance and semistandard deviation.
- You will be able to calculate these risk measures for different time periods.

Course unit 3: Models for calculating volatility

- You will be made familiar with and understand various methods of calculating moving volatilities and be able to explain them.
- You will be able to explain the differences and similarities between the EWMA, ARCH, and GARCH models and point out their respective advantages and disadvantages.
- You will perform optimizations for these procedures.

Course 2: Quantitative instruments in risk management

Course unit 1: Different types of Value at Risk and Lower Partial Moments and Extreme Value Theory

- You will calculate Value at Risk, Relative Value at Risk, and Conditional Value at Risk (Expected Shortfall) for discrete and continuous returns.
- You will be able to explain the advantages and disadvantages of using discrete and continuous returns in the Value at Risk calculation.
- You will be able to display the results graphically.
- You will be familiarized with the concepts of Lower Partial Moments and will be able to calculate and interpret them.
- You will get to know the special features of Extreme Value Theory and can differentiate these from the Value at Risk concepts and name advantages and disadvantages.
- You will be able to calculate risks using Extreme Value Theory.

Course unit 2: Determination of portfolio risks

- You will be able to calculate the portfolio risk with the variance-covariance method and place the concept in modern capital market theory.

(continued)

Course 2: Quantitative instruments in risk management (continued)

- You will have mastered portfolio risk calculation using historical simulation and will be able to explain the differences from the variance-covariance method.
- You will be able to perform Monte Carlo simulations for normally distributed and calibrated risk parameters and explain differences in the results.
- You will be familiarized with the concept of Copula functions, explaining them and using them to calculate portfolio risk. Also, you will calculate the market value of equity using stock market multiples.

Course unit 3: Hedging of hedgeable risks and modelling of non-hedgeable risks

- You will be able to identify hedgeable and non-hedgeable risks in the company.
- You will be familiar with various instruments for hedging financial risks.
- You will be able to hedge interest rate risks with futures/forwards, swaps, and options.
- You will be able to incorporate non-hedgeable risks into a company's business plan.
- You will understand the differences between planned values and expected values.
- You will aggregate individual risks in the business plan using Monte Carlo simulation.
- You will calculate different risk ratios based on the Monte Carlo simulation.

Philosophy of the Book

In this book, you will learn to apply quantitative risk management in a practical way by means of case studies. We have used our approach of financial modeling-based engineering. "Modelling means understanding." This may sound a bit "flippant," but it is the most promising way to learn about complex issues in risk management. You will see that you need some prior knowledge of mathematics and statistics for the case studies. The learning success and understanding of the models and theories are achieved through modelling.

The software used in this course is Excel. Excel is very versatile, widely used, and has relatively easy-to-learn basic functions. There are good control options and relatively large amounts of data can be processed. In addition, we use the Excel add-in "Risk Kit" in this book, which gives us further statistical and stochastic possibilities including Monte Carlo simulation. However, other software from other providers can also be used, as they are all very similar in structure and functionality. In many cases, free trial versions are available too. Participants of the "Certified Financial Engineer (CFE)" and the MBA in "Applied Quantitative Finance" receive the full version of Risk Kit for half a year for free.

We want to make our book accessible to a wide range of readers. As mentioned in the title, the reader should be able to understand step-by-step "how risk management works." Therefore, you can download the Excel spreadsheets covered in this book at:

www.certified-financial-engineer.com

These spreadsheets are based on the teaching concept of financial modelling, which consists of the following five steps:

Step 1:		The exact assignment is given here.
Step 2:		Background information on the topics is given here.
Step 3:	*fx*	Here you will get to know the respective formulas with which the result of the respective assignment can be calculated.
Step 4:		The respective implementation in Excel is shown here. At the end of each assignment, you will see an Excel screenshot of the Excel result with an overview of all the figures.
Step 5:		You will find literature references and links to the Excel tool here.

The framework of the Excel file is analogous to the Excel worksheets in the book. You can therefore first work out the contents of steps 1 and 2 yourself by reading the relevant passages in the book.

We recommend that you then transfer the formulas from the book into the downloaded Excel according to step 3.

In step 4, you then complete the cells belonging to the respective assignment in the corresponding file. Finally, you can compare your results with the Excel screenshot in the book. If you find any discrepancies, you can go back to step 2 and make changes until your Excel worksheet looks exactly like the screenshot in the book. Learning by doing!

Background Information on the Case Study "CORPORATE RISK MANAGEMENT"

Financial engineering in risk management

You are the Head of Risk Management at PHARMA GROUP and have the task of quantitatively strengthening the existing risk management framework/principles of PHARMA GROUP through the instruments of financial engineering. You are asked by the CFO of PHARMA GROUP to develop a concept for this project "Financial Engineering in Risk Management," to define the contents and to apply it to the current risk management framework/principles of PHARMA GROUP.

Part I
Course 1: Risk Analysis

After this introductory information about our case study, we now want to start with Course 1.

Course 1 covers the following topics:

- Course unit 1: Graphical presentation of risks
- Course unit 2: Variance and standard deviation
- Course unit 3: Models for calculating volatility

Quantitative Description of Risks

You, as Head of Risk Management, decide to first analyze the risks of the oil price on the basis of historical data. For this purpose, you will find the oil prices from 31.12. t(5) retrospectively for the last 5 years in the Excel file `Case Study Corporate Risk Management Part 1`. For the risk analysis, you will need the returns. You would like to know the distribution of historical oil price returns in order to be able to apply suitable risk measures. To do this, you determine the frequency distribution, distribution function, and density function of the returns.

Course Unit 1: Graphical Representation of Risks

Assignment 1: Return Calculation

Assignment

Calculate the discrete daily return and the continuous daily return for the oil price from 31.12.t(5) retrospectively for the last 5 years.

Content

The most important question in risk management is: What actually is the risk? The traditional approach is to align risk with a company's objectives. If the given goals are missed, this is negative; while if the goals are met or exceeded, this is positive. A company's goals can be divided into financial and productive goals. We will focus here on financial goals and consider risks arising from assets such as commodities, exchange rates and interest rates. The calculation of *risk* is based on mathematics and statistics and is defined as the deviation from the expected value.

Risks arise from changes in prices or values for assets. These can be measured in *absolute terms* (the price of oil has increased by US$5.00) or in *relative terms* (the price of oil has increased by 5.0%). Using relative changes allows risks of different assets to be compared and aggregated into an overall risk. The relative changes in value are called interest rates for interest-bearing financial products and returns for other financial products. We use the uniform term of *return* here. Here, we can again distinguish between discrete returns and continuous returns.

The relative change in value or *discrete return* r^d considers two individual points in time (start of investment time and the end time of the investment period) or several investment times within an investment period.

(continued)

D. Ernst, J. Häcker, *Corporate Risk Management*, Springer Texts in Business and Economics, https://doi.org/10.1007/978-3-031-53126-2_2

With a *continuous return* r^c, it is assumed that the capital invested is subject to continuous interest. The difference from the discrete return lies in the consideration of the periods in which interest is paid on the investment. It is quite possible that an investment will earn interest not only monthly, but also weekly, daily or even hourly, or at even shorter intervals. The smaller the interest periods, the smaller the difference between the discrete and the continuous returns.

In risk management, the question always arises as to whether we want to *use discrete or continuous returns* as the basis for our further calculations. In the following, the decision to use discrete or continuous returns depends on the available database. If we work with empirical data and empirical distributions, it is a good idea to calculate the relevant risk parameters with the intuitively understandable discrete return. This also applies to historical simulation. If, on the other hand, we want to perform risk calculations on the basis of normal distributions, we opt for continuous returns, as normal distributions with continuous returns can be modelled better. This also applies to the calculation of portfolio risk using the variance-covariance method and Monte Carlo simulation, in which a normal distribution of the risk factors and the portfolio is assumed and a separate density function is, respectively, created by calibration. Furthermore, we use continuous returns in modeling stochastic processes.

Important Formulas

Calculation of the *discrete daily return*:

$$\boxed{r_t^d = \frac{V_t}{V_{t-1}} - 1 = \frac{V_t - V_{t-1}}{V_{t-1}}} \tag{1}$$

r_t^d = Discrete return at time t, here on the day t
V_t = Value at time t, here on day t
V_{t-1} = Value at time $t - 1$

Excel workbook: `Case Study Corporate Risk Management Part 1` ➲ Worksheet: `Returns`

Excel example: `D8=(C8/C7)-1`

Calculation of the *continuous daily return*:

$$\boxed{r^c = \ln\left(\frac{V_t}{V_{t-1}}\right)} \tag{2}$$

r^c = Continuous return
V_t = Value at time t, here on day t
V_{t-1} = Value at time $t-1$

Excel example: `E8=LN(C8/C7)`

Procedure

- Create a column for the oil price (column `C`). Link the cells in this column to the values from the worksheet `Assumptions Oil Price` so that the oil prices for the specified period are displayed on the worksheet `Returns`.
- Calculate the discrete daily return using the formula above `D8=(C8/C7)-1`.
- Then calculate the continuous daily return according to the formula above `E8=LN(C8/C7)`.

Result (Fig. 1)

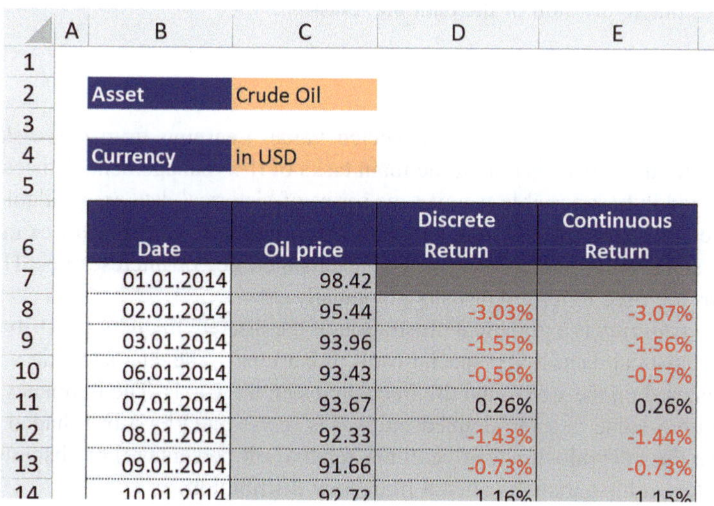

	A	B	C	D	E
1					
2		Asset	Crude Oil		
3					
4		Currency	in USD		
5					
6		Date	Oil price	Discrete Return	Continuous Return
7		01.01.2014	98.42		
8		02.01.2014	95.44	-3.03%	-3.07%
9		03.01.2014	93.96	-1.55%	-1.56%
10		06.01.2014	93.43	-0.56%	-0.57%
11		07.01.2014	93.67	0.26%	0.26%
12		08.01.2014	92.33	-1.43%	-1.44%
13		09.01.2014	91.66	-0.73%	-0.73%
14		10.01.2014	92.72	1.16%	1.15%

Fig. 1 Discrete and continuous returns

Literature and References to the Excel Tool

Ernst, D. (2020): Corporate Risk Management, Course 1—Assignment 1.

Standard literature:

Hopkin, P. (2018): Fundamentals of Risk Management: Understanding, Evaluating and Implementing Effective Risk, 5. Edition, New York.

Hull, J.C. (2018): Risk Management and Financial Institutions, 5. Edition, Hoboken (New Jersey).

See Excel file `Case Study Risk Management Part 1`, Excel worksheet `Returns`

Assignment 2: Creating a Histogram

Assignment

Create a histogram for the discrete daily returns of the oil price from 31.12.t (5) retrospectively for the last 5 years to show the frequency distribution graphically. Select a suitable division of the data into classes.

Content

Risk is the deviation from an expected value. Learning from past risks to evaluate future risks is one of the main tasks of risk management. This can be achieved in best possible way if time series of historical data are available for longer time periods. This can be exhibited in practical scenario, for example, with stock prices, commodity prices, exchange rates or interest rates. Histograms are used to represent risks graphically.

A *histogram* is a graphical illustration of the discrete frequency distribution of a statistical data. It is a special form of bar chart. The characteristic values are entered on the *x*-axis and the frequencies on the *y*-axis. The frequency of a measured value in a pre-defined interval is represented by a bar-shaped area above the interval—this can be either relative (in percentage) or absolute. In statistics, a histogram is called a frequency distribution.

A histogram provides a graphical overview of the distribution of returns. This makes it possible to illustrate the size of the spread and the risk of the asset. A histogram allows larger volumes of data to be captured compared with a simple table. Lastly, any concentration of extreme risks at the edges of the distribution can be quickly identified.

Important Formulas

Workbook: `Case Study Corporate Risk Management Part 1` ➲ Worksheet: `Histogram`

Determination of the *minimum* rate of the return:

Excel example: `H8=MIN(D8:D1310)`

Determination of the *maximum* of all the returns:

Excel example: `H9=MAX(D8:D1310)`

Determination of the *mean* value of the returns:

Excel example: `H10=AVERAGE(D8:D1310)`

Determination of the number of returns:

Excel example: `H11=COUNT(D8:D1310)`

Procedure

- For creating a histogram, the discrete daily returns are first calculated in column `D`.
- Then, the calculations are made using the respective Excel functions:
  - The minimum `H8=MIN(D8:D1310)`,
  - The maximum `H9=MAX(D8:D1310)`,
  - The mean `H10=AVERAGE(D8:D1310)`
  - The number of returns `H11=COUNT(D8:D1310)`
- These help to define suitable intervals (class ranges) for the histogram.
- To generate the histogram, the class range is defined here in cells `F14:F40`. This information is taken from the worksheet `Assumptions General`.
- In our Excel example, the classes range from −13% to 13% and increase by 1.0% steps.
- Using the `HISTOGRAM` analysis function, the distribution can be determined very easily.
- You can access the function in Excel via `Data` ➲ `Analysis` ➲ `Data Analysis` ➲ `Histogram`. In case `Analysis` is not activated in your Excel version, go to Options ➲ `Add-Ins` ➲ `Go ...` and check `Analysis ToolPak` and preferably also `Solver Add-in`, which we will need later. Confirm your selection with `OK`.

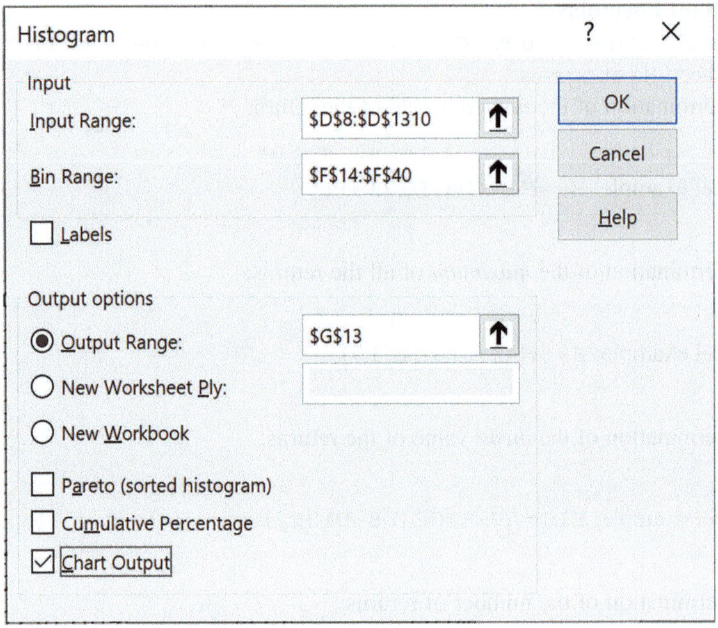

Fig. 2 Inputs for creating a histogram

- Figure 2 shows the input pane in Excel for creating the histogram.
- The discrete daily returns cells `D8:D1310` are entered as the input range, the previously determined class upper limits in cells `F14:F40` as the class range, and cell `G13`, from which the result is displayed, as the output range.
- In addition, the field `Chart Output` is clicked to immediately obtain a diagram from the data.
- The columns Class and Frequency are automatically inserted and calculated.
- From a visual perspective, it is recommended that you adapt the chart in Excel to your own design later on.

Result (Fig. 3)

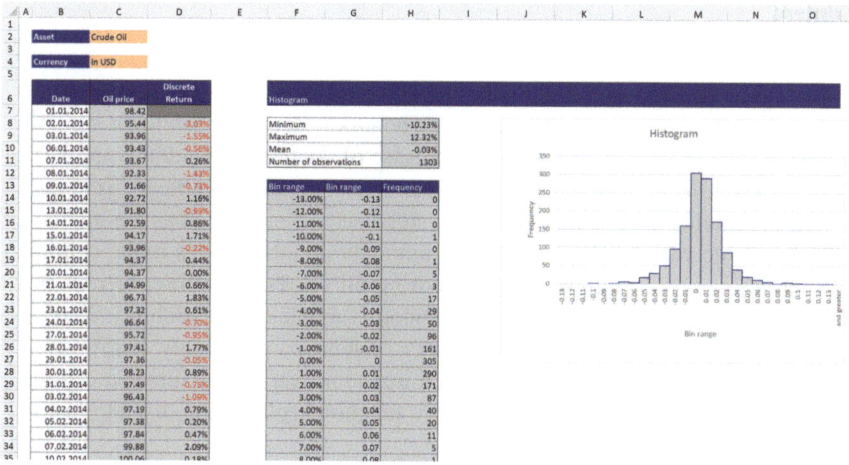

Fig. 3 Creating a histogram

Literature and References to the Excel Tool

Ernst, D. (2020): Corporate Risk Management, Course 1—Assignment 2.

Standard literature:

Hopkin, P. (2018): Fundamentals of Risk Management: Understanding, Evaluating and Implementing Effective Risk, 5. Edition, New York.

Hull, J.C. (2018): Risk Management and Financial Institutions, 5. Edition, Hoboken (New Jersey).

See Excel file `Case Study Risk Management Part 1`, Excel worksheet `Histogram`

Assignment 3: Creating a Density Function and a Distribution Function

Assignment

Create a density function and a distribution function for the continuous daily returns of the oil price based on the assumption of a normal distribution.

Content

For a *discrete* random variable, there are *finite* possible observed values, each of which has a positive probability. Therefore, these probabilities can be assigned to the observed values. In the case of *continuous* random variables, on the other hand, there are *infinitely* many theoretically possible realizations. The probability with which a certain value occurs—calculated as the number of favorable values by the number of infinitely possible values in the continuous case—is accordingly equal to zero for all values. Therefore, there is no probability function for continuous random variables. In probability theory, the *density function* is a tool to describe a continuous probability distribution. It describes the probability with which a random variable assumes a certain characteristic value. Hence, the density function replaces the discrete histogram function discussed above.

In this case, the density function takes its place. A density function indicates how densely the variables under consideration are distributed around an arbitrary point. The higher the density at this point, the higher the probability that a variable from this area is realized.

A *distribution function* describes the relationship between a random variable and its probabilities. It indicates the probability with which a random variable assumes at most a certain range of values. The probability function thus describes the cumulative probability. The density function is the first derivative of the distribution function.

One of the most important continuous probability distributions is the normal distribution. The density function of the normal distribution has a bell-shaped appearance. The appearance and properties of the normal distribution are determined by two parameters:

- Expected value μ: It determines at which point the normal distribution has its maximum.
- Standard deviation σ: It shows the dispersion around the expected value.

The total area enclosed by the curve of normal distribution (hence the integral from $-\infty$ to ∞) is always 1.

The *standard normal distribution* is a normal distribution in which the expected value $= 0$ and the standard deviation $= 1$.

Important Formulas

Calculation of the *expected value*, which corresponds to the *mean* of the historical return:

$$\mu = \frac{1}{m} \sum_{t=1}^{m} r_t \tag{3}$$

μ = Expected rate of return
m = Number of observations
r_t = Return at time t

Workbook: `Case Study Corporate Risk Management Part 1` ➲ Worksheet: `Density & distribution function`

Excel example: `E7=AVERAGE(Returns!E8:E1310)`

Calculation of the *standard deviation* (based on a sample) of continuous daily returns:

$$\sigma = \sqrt{\frac{1}{m-1} \sum_{t=1}^{m} (r_t - \mu)^2} \tag{4}$$

σ = Standard deviation of returns
m = Number of observations
r_t = Return at time t
μ = Expected rate of return

Excel example: `E8=STDEV.S(Returns!E8:E1310)`

Determination of the *probability* of returns:

Excel example: `B11 ='Assumptions General'!C37`

Calculation of the *quantiles* of the normal distribution:

Excel example: `C11=NORM.INV(B11;$E$7;$E$8)`

Interpretation:

- 0.01% of returns are less than −8.44%.
- 50.00% of returns are less than −0.06%.

Calculation of the *density* of the normal distribution:

Excel example: `D11=NORM.DIST(C11;$E$7;$E$8;FALSE)`

Interpretation: There is no economic interpretation for the density.
Calculation of the *cumulative probability*:

Excel example: `E11=NORM.DIST(C11;$E$7;$E$8;TRUE)`

Interpretation:

- The probability that the continuous daily return will not exceed −8.44% is 0.01%.
- The probability that the continuous daily return will not exceed −0.06% is 50.00%.

Procedure

- Create a column for the probabilities (column `B`). Link the cells in this column to the values in the worksheet `Assumptions General` so that the probabilities are displayed on the worksheet `Density & Distribution Function`.
- Calculate the values of the normal distribution `C11=NORM.INV(B11;$E$7;$E$8)`, the probability density `D11=NORM.DIST(C11;$E$7;$E$8;FALSE)` and the cumulative probability according to the formula `E11=NORM.DIST(C11;$E$7;$E$8;TRUE)`.

Result (Figs. 4 and 5)

	A	B	C	D	E
4		Currency	in USD		
5					
6		Assumptions			
7		Mean			-0.0594%
8		Standard deviation			2.2540%
9					
10		Probability	Normal distribution	Density	Cumulative probability
11		0.0001	-0.0844	0.0176	0.0001
12		0.0002	-0.0804	0.0336	0.0002
13		0.0003	-0.0779	0.0491	0.0003
14		0.0004	-0.0762	0.0641	0.0004
15		0.0005	-0.0748	0.0788	0.0005
16		0.0006	-0.0736	0.0933	0.0006
17		0.0007	-0.0726	0.1076	0.0007
18		0.0008	-0.0717	0.1217	0.0008
19		0.0009	-0.0710	0.1356	0.0009
20		0.0010	-0.0702	0.1494	0.0010
21		0.0020	-0.0655	0.2813	0.0020
22		0.0030	-0.0625	0.4059	0.0030
23		0.0040	-0.0604	0.5256	0.0040
24		0.0050	-0.0587	0.6415	0.0050

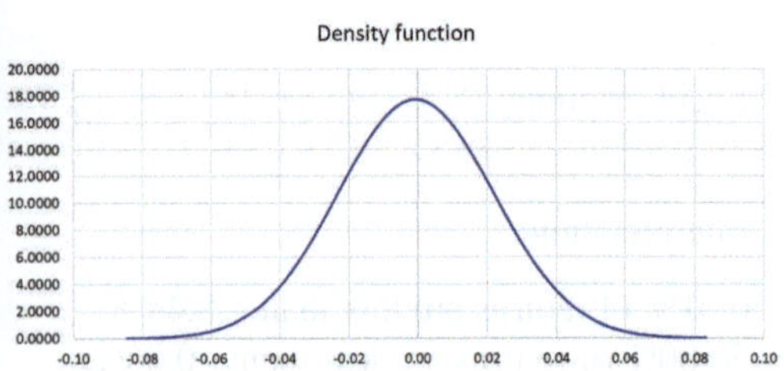

Fig. 4 Creation of a density function

25	0.0060	-0.0572	0.7543	0.0060
26	0.0070	-0.0560	0.8645	0.0070
27	0.0080	-0.0549	0.9725	0.0080
28	0.0090	-0.0539	1.0784	0.0090
29	0.0100	-0.0530	1.1824	0.0100
30	0.0200	-0.0469	2.1481	0.0200
31	0.0300	-0.0430	3.0187	0.0300
32	0.0400	-0.0401	3.8231	0.0400
33	0.0500	-0.0377	4.5756	0.0500
34	0.0600	-0.0356	5.2849	0.0600
35	0.0700	-0.0339	5.9568	0.0700
36	0.0800	-0.0323	6.5956	0.0800
37	0.0900	-0.0308	7.2045	0.0900
38	0.1000	-0.0295	7.7860	0.1000
39	0.1100	-0.0282	8.3423	0.1100
40	0.1200	-0.0271	8.8749	0.1200
41	0.1300	-0.0260	9.3853	0.1300
42	0.1400	-0.0249	9.8747	0.1400

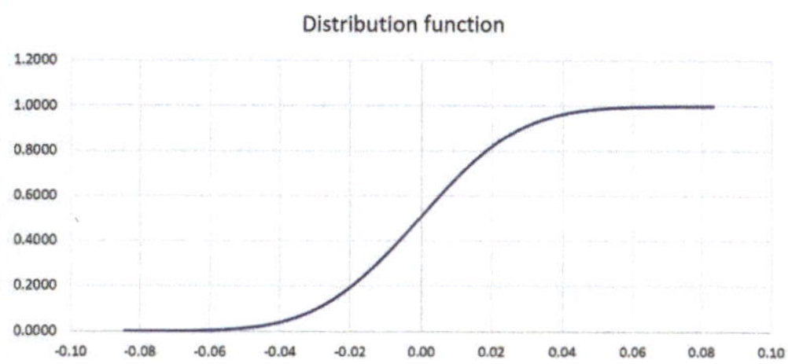

Fig. 5 Creation of a distribution function

Literature and References to the Excel Tool

Ernst, D. (2020): Corporate Risk Management, Course 1—Assignment 3.

Standard literature:

Hopkin, P. (2018): Fundamentals of Risk Management: Understanding, Evaluating and Implementing Effective Risk, 5. Edition, New York.

Hull, J.C. (2018): Risk Management and Financial Institutions, 5. Edition, Hoboken (New Jersey).

See Excel file Case Study `Corporate Risk Management`, Excel worksheet `Density & Distribution Function`

Assignment 4: Calculation of the Skewness

Assignment

Calculate the skewness for the discrete daily oil price returns.

Interpret the result and compare your statement with the histogram.

Content

Skewness and kurtosis are measures that describe the deviation of a distribution from the normal distribution. The excess indicates the difference of the kurtosis of the considered function to the kurtosis of the density function of a normally distributed random variable, which has a kurtosis of three. For normally distributed values, both skewness and excess are considered as zero. The further away the values are from zero, the less likely the data are not normally distributed.

(continued)

The skewness indicates whether the distribution is symmetrical or not. *Skewness* is a statistical measure that describes the nature and strength of the asymmetry of a probability distribution. It indicates whether and how strongly the distribution is inclined to the right (right skew = positive skewness) or to the left (left skew = negative skewness). Any distribution that is not symmetrical is called a skew.

The skewness can assume any real value:

- Negative skewness, skewness <0, is called left- or negatively-skewed distribution. In typical cases, it falls more flatly on the left side than on the right.
- A positive skewness, skewness >0, is called a right- or positively-skewed distribution. Typically, it falls more flatly on the right side than on the left.

Important Formulas

The following formulas are used to calculate the skewness. A sample is assumed as the data basis:

$$\boxed{\text{Skewness} = \frac{m_3}{\sigma^3}} \tag{5}$$

m_3 = Central moment third order
σ^3 = Third power of the standard deviation

The central moment of the third order is divided by the third power of the standard deviation.

$$\boxed{m_3 = \frac{m}{(m-1)\cdot(m-2)} \sum_{t=1}^{m} (r_t - \overline{r_t})^3} \tag{6}$$

$$\boxed{\sigma^3 = \left(\sqrt{\frac{1}{m-1} \sum_{t=1}^{m} (r_t - \overline{r_t})^2}\right)^3} \tag{7}$$

With the "correction factor" $\frac{m}{(m-1)\cdot(m-2)}$, the estimator for the skewness will be unbiased.

Workbook: `Case Study Corporate Risk Management Part 1` ➲ Worksheet: `Skewness & Kurtosis`

Calculation of the skewness:

Excel example: K18=K16/K17

Calculation of the skewness using the Excel function:

Excel example: K19=SKEW(D8:D1310)

Procedure

- The skewness can be calculated manually and with the Excel function SKEW.
- When using the Excel function SKEW, only the returns need to be entered. The disadvantage of the Excel function SKEW is that it is a black box. To begin with, one does not know in this Excel function how the calculation is carried out and whether a sample or a population of the data is assumed.
- If the Excel function SKEW is used, first the central moment of third order is calculated assuming a sample as a base. For this purpose, first, the difference between the returns and their mean value is calculated for all returns (D8:D1310). The results are raised to the power of three (F8:F1310). Then the sum of the third power is calculated from these results K11=SUM(F8:F1310). In the next step, we determine the correction factor for a finite population K15=COUNT(D8:D1310)/((COUNT(D8:D1310)-1) *(COUNT(D8:D1310)-2)). If the sum of the third power (K11) is multiplied by the correction factor (K15), the result is the third-order central moment (K16=K15*K11).
- The third power of the standard deviation is then calculated, also assuming a sample (K17=STDEV.S(D8:D1310)^3).
- Dividing the central moment of the third order by the third power of the standard deviation results in the skewness K18=K16/K17.
- The skewness of 0.233804207 indicates that the distribution at hand has a slightly right-skewed shape. Overall, however, it can be described as very symmetrical. This is also made clear by Fig. 6.

Result (Figs. 6 and 7)

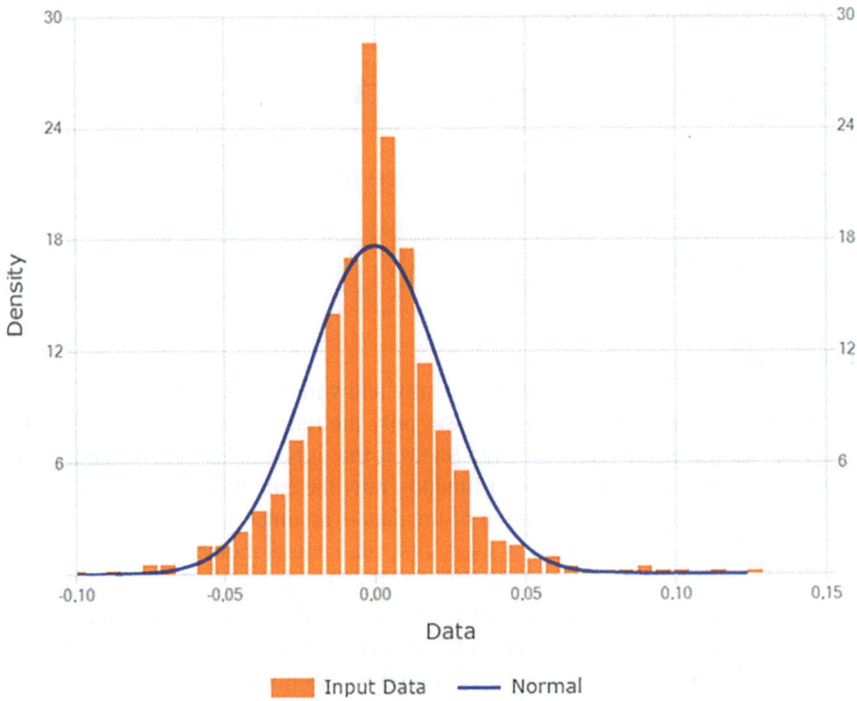

Fig. 6 Comparison of the histogram and the normal distribution using skewness

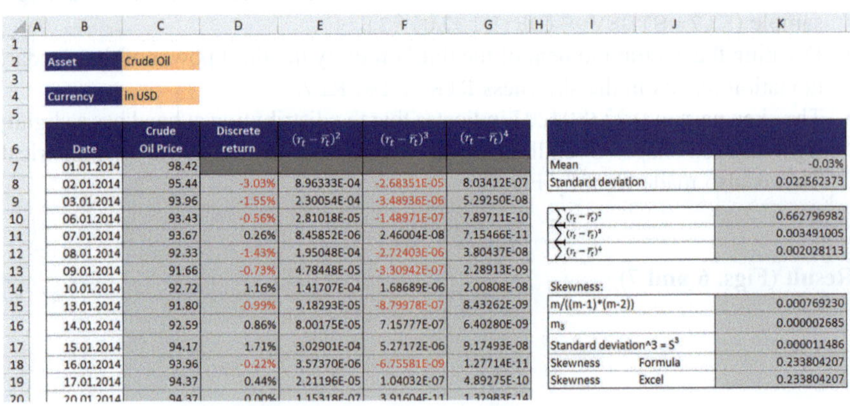

Asset	Crude Oil
Currency	in USD

Date	Crude Oil Price	Discrete return	$(r_t - \bar{r}_t)^2$	$(r_t - \bar{r}_t)^3$	$(r_t - \bar{r}_t)^4$
01.01.2014	98.42				
02.01.2014	95.44	-3.03%	8.96333E-04	-2.68351E-05	8.03412E-07
03.01.2014	93.96	-1.55%	2.30054E-04	-3.48936E-06	5.29250E-08
06.01.2014	93.43	-0.56%	2.81018E-05	-1.48971E-07	7.89711E-10
07.01.2014	93.67	0.26%	8.45852E-06	2.46004E-08	7.15466E-11
08.01.2014	92.33	-1.43%	1.95048E-04	-2.72403E-06	3.80437E-08
09.01.2014	91.66	-0.73%	4.78448E-05	-3.30942E-07	2.28913E-09
10.01.2014	92.72	1.16%	1.41707E-04	1.68689E-06	2.00808E-08
13.01.2014	91.80	-0.99%	9.18293E-05	-8.79978E-07	8.43262E-09
14.01.2014	92.59	0.86%	8.00175E-05	7.15777E-07	6.40280E-09
15.01.2014	94.17	1.71%	3.02901E-04	5.27172E-06	9.17493E-08
16.01.2014	93.96	-0.22%	3.57370E-06	-6.75581E-09	1.27714E-11
17.01.2014	94.37	0.44%	2.21196E-05	1.04032E-07	4.89275E-10
20.01.2014	94.37	0.00%	1.15318E-07	3.91604E-11	1.32983E-14

Mean	-0.03%
Standard deviation	0.022562373
$\sum (r_t - \bar{r}_t)^2$	0.662796982
$\sum (r_t - \bar{r}_t)^3$	0.003491005
$\sum (r_t - \bar{r}_t)^4$	0.002028113
Skewness:	
m/((m-1)*(m-2))	0.000769230
m_3	0.000002685
Standard deviation^3 = S^3	0.000011486
Skewness Formula	0.233804207
Skewness Excel	0.233804207

Fig. 7 Calculation of the skewness

Literature and References to the Excel Tool

Ernst, D. (2020): Corporate Risk Management, Course 1—Assignment 4.
Standard literature:
Hopkin, P. (2018): Fundamentals of Risk Management: Understanding, Evaluating and Implementing Effective Risk, 5. Edition, New York.
Hull, J.C. (2018): Risk Management and Financial Institutions, 5. Edition, Hoboken (New Jersey).
See Excel file `Case Study Corporate Risk Management Part 1`, Excel worksheet `Skewness & Kurtosis`

Assignment 5: Calculation of the Kurtosis

Assignment
Calculate the kurtosis for the discrete daily returns of the oil price.
Interpret the result and compare your statement with the histogram.

Content

The *kurtosis* describes whether the distribution is peaked or flattened in contrast to the normal distribution. Kurtosis is a measure of the steepness or "peakedness" of a (single-peak) probability distribution. The excess indicates the difference between the kurtosis of the function under consideration and the kurtosis of the density function of a normally distributed random variable. If the kurtosis of the function under consideration corresponds to the kurtosis of a normal distribution, the excess is zero.

- A positive excess indicates a pointed distribution (also called leptokurtic distribution).
- A negative excess indicates a flat distribution (platykurtic distribution).

A peaked distribution has a positive excess. Here, more observations than usual are located in the edges of the distribution, which is why it is also called heavy tailed.

A negative excess describes a flattened distribution. Such a distribution has thin ends (light tailed) compared to the normal distribution.

Important Formulas
The following formulas form the basis for calculating the kurtosis:

$$\boxed{\text{Kurtosis} = \frac{m_4}{\sigma^4}} \tag{8}$$

m_4 = Central moment fourth order
σ^4 = Fourth power of the standard deviation

The central moment of the fourth order is divided by the fourth power of the standard deviation.

The following formula is the basis for the calculation of the excess:

$$\boxed{\text{excess} = \text{kurtosis} - \text{kurtosis (normal distribution)}} \tag{9}$$

If you subtract the kurtosis of the normal distribution from the calculated kurtosis of the function under consideration, you get the excess. The kurtosis of a normal distribution is always 3. With the help of the excess, it can be determined as to what extent the curvature of a distribution resembles the curvature of the known normal distribution.

The formula for the fourth-order central moment is:

$$\boxed{m_4 = \frac{m \cdot (m+1)}{(m-1) \cdot (m-2) \cdot (m-3)} \sum_{t=1}^{m} (r_t - \overline{r_t})^4} \tag{10}$$

With the "correction factor" $\frac{m \cdot (m+1)}{(m-1) \cdot (m-2) \cdot (m-3)}$ the estimator for kurtosis becomes unbiased.

$$\boxed{\sigma^4 = \left(\sqrt{\frac{1}{m} \sum_{t=1}^{m} (r_t - \overline{r_t})^2}\right)^4} \tag{11}$$

Also, the kurtosis of the normal distribution having the value of 3 still has to be adjusted in the above as follows so that the results will be unbiased.

$$\boxed{3 \cdot \frac{(m-1)^2}{(m-2) \cdot (m-3)}} \tag{12}$$

Workbook: `Case Study Corporate Risk Management Part 1` ➲ Worksheet: `Skewness & Kurtosis`.

Calculation of the kurtosis:

Excel example: `K26=K24/K25`

Calculation of the excess:

Excel example: `K27=K24/K25-K23`

Calculation of the excess using the Excel function:

Excel example: `K28=KURT(D8:D1310)`

Procedure

- The kurtosis can be calculated manually and with the Excel function `KURT`.
- When using the Excel function `KURT`, only the returns need to be entered. The disadvantage of the Excel function `KURT` is that it is a black box. Hence, you do not know how the calculation is performed and whether a sample or a population of the data is assumed.
- If the Excel function `KURT` is reproduced, first the fourth-order central moment is calculated assuming a sample as a base. For this purpose, the difference between the returns and their mean value is first calculated for all returns (`D8:D1310`). The results are raised to the power of four (`G8:G1310`). Then the sum of this fourth power is calculated from these results `K12=SUM(G8:G1310)`. In the next step, we determine the correction factor for a finite population `K22=(COUNT(D8:D1310)*(COUNT(D8:D1310)+1))/((COUNT(D8:D1310)-1)*(COUNT(D8:D1310)-2)*(COUNT(D8:D1310)-3))`. If the sum of the fourth power (`K12`) is multiplied by the correction factor (`K22`), the result is central moment of fourth order (`K24=K22*K12`).
- Then, the fourth power of the standard deviation is also calculated assuming a sample `K25=STDEV.S(D8:D1310)^4`.
- Dividing the central moment of fourth order by the fourth power of the standard deviation, the kurtosis results are obtained (`K26=K24/K25`).
- After subtracting the corrected kurtosis of the normal distribution, one obtains the excess, which can be interpreted as indicated above `K27=K24/K25-K23`.
- We get the same result from the Excel function KURT, although this actually calculates the excess and not the kurtosis. It is therefore always worthwhile to follow the Excel functions critically.
- A positive excess of 3.031774784 indicates a tapered distribution (also called platykurtic distribution). This is also made clear by Fig. 8.

Result (Figs. 8 and 9)

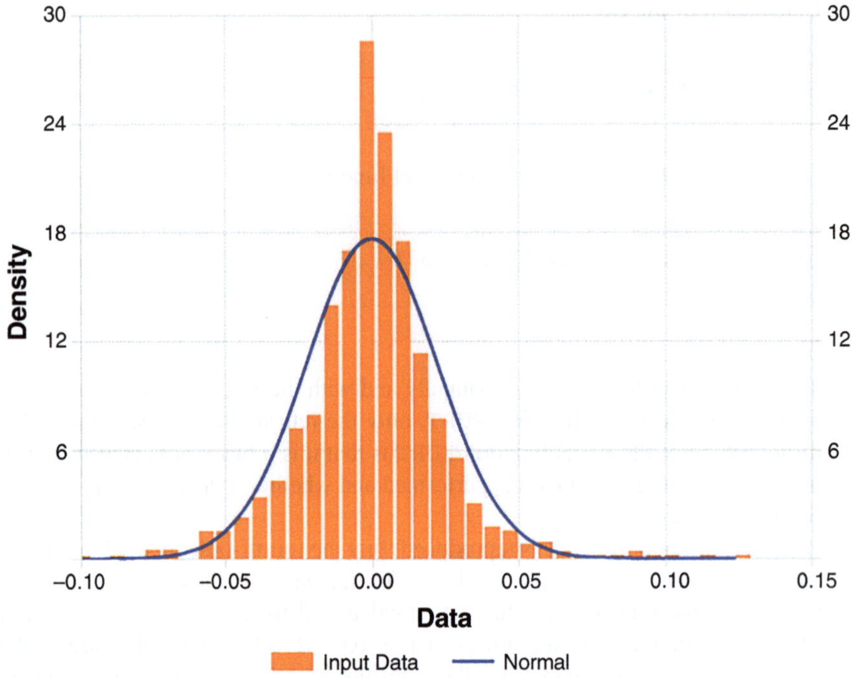

Fig. 8 Comparison of the histogram and the normal distribution using kurtosis

	A	B	C	D	E	F	G	H	I	J	K
21		21.01.2014	94.99	0.66%	4.77408E-05	3.29863E-07	2.27918E-09		Kurtosis:		
22		22.01.2014	96.73	1.83%	3.48095E-04	6.49451E-06	1.21170E-07		(m*(m+1))/((m-1)*(m-2)*(m-3))		0.000771597
23		23.01.2014	97.32	0.61%	4.14612E-05	2.66970E-07	1.71903E-09		(3*(m-1)^2)/((m-2)*(m-3))		3.006924851
24		24.01.2014	96.64	-0.70%	4.41916E-05	-2.93771E-07	1.95289E-09		m_4		0.000001565
25		27.01.2014	95.72	-0.95%	8.42776E-05	-7.73692E-07	7.10271E-09		Standard deviation^4 = S^4		0.000000259
26		28.01.2014	97.41	1.77%	3.23829E-04	5.82738E-06	1.04865E-07		Kurtosis	Formel	6.038699635
27		29.01.2014	97.36	-0.05%	3.01748E-08	-5.24162E-12	9.10516E-16		Excess	Formula	3.031774784
28		30.01.2014	98.23	0.89%	8.60348E-05	7.98015E-07	7.40198E-09		Excess	Excel (Excess)	3.031774784
29		31.01.2014	97.40	-0.75%	[illegible]	[illegible]	[illegible]				

Fig. 9 Calculating the kurtosis

Literature and References to the Excel Tool

Ernst, D. (2020): Corporate Risk Management, Course 1—Assignment 5.

Standard literature:

Hopkin, P. (2018): Fundamentals of Risk Management: Understanding, Evaluating and Implementing Effective Risk, 5. Edition, New York.

Hull, J.C. (2018): Risk Management and Financial Institutions, 5. Edition, Hoboken (New Jersey).

See Excel file `Case Study Corporate Risk Management Part 1`, Excel worksheet `Skewness & Kurtosis`

Course Unit 2: Variance and Standard Deviation

Assignment 6: Calculation of the Variance

Assignment

Calculate the variance for the continuous daily returns of the oil price based on the assumption of a sample.

Content

> The *variance* is a measure of dispersion that characterizes the distribution of values around the mean value.

D. Ernst, J. Häcker, *Corporate Risk Management*, Springer Texts in Business and Economics, https://doi.org/10.1007/978-3-031-53126-2_3

Important Formulas

Calculation of the variance (based on a sample) of continuous daily returns:

$$\boxed{\sigma^2 = \frac{1}{m-1}\sum_{t=1}^{m}(r_t-\mu)^2} \tag{1}$$

σ^2 = Variance of returns
m = Number of observations
r_t = Return at time t
μ = Expected rate of return

Workbook: `Case Study Corporate Risk Management Part 1` ➲ Worksheet: `Variance & Standard Deviation`

> Excel example: `I8=VAR.S(D8:D1310)`

Procedure

- Calculate the variance based on the continuous daily returns in column `D`.

Result (Fig. 1)

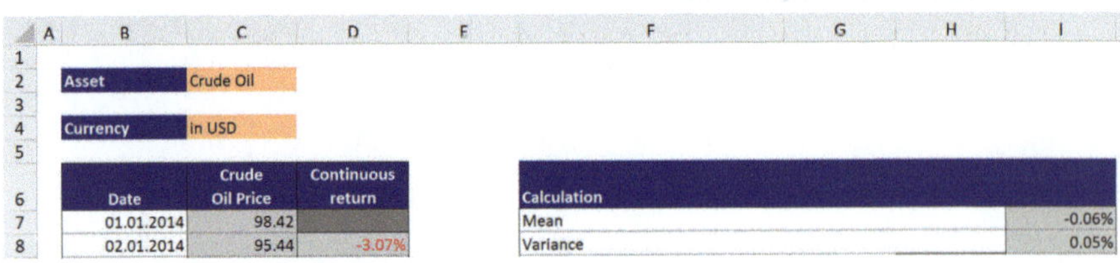

Fig. 1 Calculation of the variance

Literature and References to the Excel Tool

Ernst, D. (2020): Corporate Risk Management, Course 1—Assignment 6.

Standard literature:

Hopkin, P. (2018): Fundamentals of Risk Management: Understanding, Evaluating and Implementing Effective Risk, 5. Edition, New York.

Hull, J.C. (2018): Risk Management and Financial Institutions, 5. Edition, Hoboken (New Jersey).

See Excel file `Case Study Corporate Risk Management Part 1`, Excel worksheet `Variance & Standard Deviation`

Assignment 7: Calculation of the Annualized and Sub-periodic Variance

Assignment

From the daily variance, calculate the annualized variance for 250 trading days. Then, calculate the monthly variance from the annualized variance.

Content

The *comparison of risks* only makes sense if the same periods are considered across. For example, it makes no sense to compare variances at a monthly level with variances at an annual level. For this reason, risks should be annualized.

When calculating the variance and the standard deviation, continuous returns are preferred as a basis for calculation. Compared to discrete returns, continuous returns have the property and advantage of time additivity. This means that with continuous returns, the return over a longer period can be calculated as the sum of the returns over the short periods.

Important Formulas

If the variance for a period of length t is given (e.g., the daily variance), the following applies to the variance of a period comprising $n \cdot t$ time units (e.g., the annualized variance):

$$\boxed{\sigma^2_{\text{annualized}} = \sigma^2_t \cdot m} \tag{2}$$

$\sigma^2_{\text{annualized}}$ = Variance of return at annual level
σ^2_t = Intra-year variance of return

t = Intra-year period
m = Number of periods

Workbook: `Case Study Corporate Risk Management Part 1` ➲ Worksheet: `Variance & Standard Deviation`

Excel example: `I9=I8*H9`

It applies to the conversion of the variance of the return at the annual level into a variance of the return during the year:

$$\sigma_t^2 = \sigma_{\text{annualized}}^2 \cdot \frac{1}{m} = \frac{\sigma_{\text{annualized}}^2}{m} \tag{3}$$

Excel example: `I10=I9/H10`

Procedure

- Calculate the annualized and monthly variance based on the above formulas.

Result (Fig. 2)

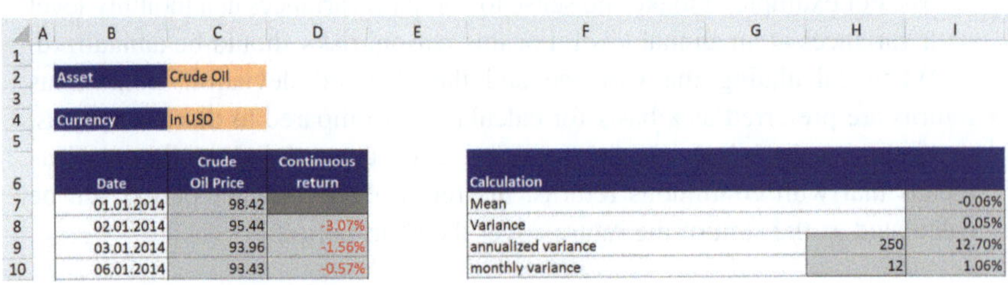

Fig. 2 Calculation of annualized and intra-year variance

Literature and References to the Excel Tool

Ernst, D. (2020): Corporate Risk Management, Course 1—Assignment 7.
Standard literature:
Hopkin, P. (2018): Fundamentals of Risk Management: Understanding, Evaluating and Implementing Effective Risk, 5. Edition, New York.
Hull, J.C. (2018): Risk Management and Financial Institutions, 5. Edition, Hoboken (New Jersey).
See Excel file `Case Study Corporate Risk Management Part 1`, Excel worksheet `Variance & Standard Deviation`

Assignment 8: Calculation of the Standard Deviation

Assignment

Calculate the standard deviation for the continuous daily returns of the oil price based on the assumption of a sample.

Content

The *standard deviation* is the average distance of all the measured values of a variable from the mean. The standard deviation can be calculated either directly or as the square root of the variance.

Important Formulas

Calculation of the standard deviation (based on a sample) of discrete daily returns:

$$\boxed{\sigma = \sqrt{\frac{1}{m-1}\sum_{t=1}^{m}(r_t-\mu)^2}} \tag{4}$$

σ = Standard deviation of return
m = Number of observations
r_t = Return at time t
μ = Expected rate of return

Workbook: `Case Study Corporate Risk Management Part 1` ➲ Worksheet: `Variance & Standard Deviation`

Excel example: `I11=STDEV.S(D8:D1310)`

Calculation of the standard deviation (based on a sample) as the square root of the variance:

$$\boxed{\sigma = \sqrt{\sigma^2} = \sqrt{\mathrm{Var}[r]}} \tag{5}$$

σ = Standard deviation of returns
σ^2 = Variance

Excel example: `I12=SQRT(I8)`

Procedure

- Calculate the standard deviation directly.
- First calculate the variance and then derive the square root.

Result (Fig. 3)

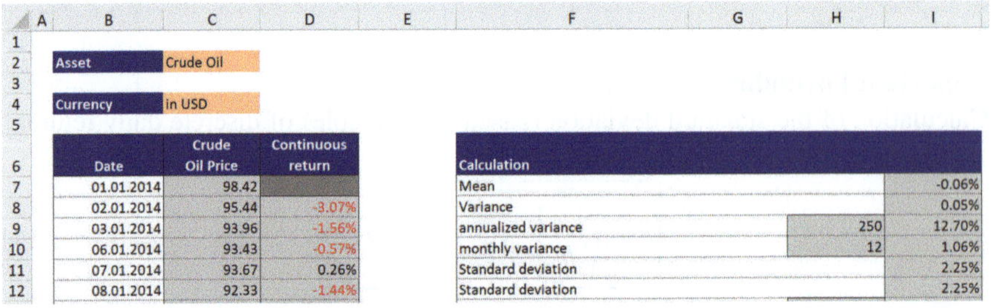

	A	B	C	D
1				
2		Asset	Crude Oil	
3				
4		Currency	in USD	
5				
6		Date	Crude Oil Price	Continuous return
7		01.01.2014	98.42	
8		02.01.2014	95.44	-3.07%
9		03.01.2014	93.96	-1.56%
10		06.01.2014	93.43	-0.57%
11		07.01.2014	93.67	0.26%
12		08.01.2014	92.33	-1.44%

	F	H	I
6	Calculation		
7	Mean		-0.06%
8	Variance		0.05%
9	annualized variance	250	12.70%
10	monthly variance	12	1.06%
11	Standard deviation		2.25%
12	Standard deviation		2.25%

Fig. 3 Calculation of the standard deviation

Literature and References to the Excel Tool

Ernst, D. (2020): Corporate Risk Management, Course 1—Assignment 8.
Standard literature:

Hopkin, P. (2018): Fundamentals of Risk Management: Understanding, Evaluating and Implementing Effective Risk, 5. Edition, New York.

Hull, J.C. (2018): Risk Management and Financial Institutions, 5. Edition, Hoboken (New Jersey).

See Excel file `Case Study Corporate Risk Management Part 1`, Excel worksheet `Variance & Standard Deviation`

Assignment 9: Calculation of the Annualized and Sub-periodic Standard Deviation

Assignment

Calculate the annualized standard deviation for 250 trading days from the daily variance. Then calculate the monthly standard deviation from the annualized standard deviation.

Content

It is also necessary to consider the same time periods when *comparing the standard deviation*. For example, only annualized standard deviations must be compared with other annualized standard deviations. For this reason, the risk values for different time periods should be made comparable.

Important Formulas

The standard deviation is annualized as follows:

$$\boxed{\sigma_{\text{annualized}} = \sigma_t \cdot \sqrt{m}} \qquad (6)$$

$\sigma_{\text{annualized}}$ = Standard deviation of returns at annual level
σ_t = Intra-year standard deviation of return
t = Single intra-year period
m = Number of periods

Workbook: `Case Study Corporate Risk Management Part 1` ➲ Worksheet: `Variance & Standard Deviation`

> Excel example: `I13=I11*SQRT(H13)`

It applies to the conversion of the standard deviation of the returns at annual level into a standard deviation of the returns during the year:

$$\boxed{\sigma_t = \sigma_{\text{annualized}} \cdot \frac{1}{\sqrt{m}} = \frac{\sigma_{\text{annualized}}}{\sqrt{m}}} \tag{7}$$

> Excel example: `I14=I13/SQRT(H14)`

Procedure

- Calculate the annualized and monthly standard deviation based on the above formulas.
- The annualized standard deviation is particularly relevant. It amounts to 35.64%. If we assume that the volatility (standard deviation) on the stock market is approximately 20%, we can see that the oil prices have significantly higher risk levels in comparison to even stock markets (at least for the period under consideration).

Result (Fig. 4)

Asset	Crude Oil
Currency	in USD

Date	Crude Oil Price	Continuous return
01.01.2014	98.42	
02.01.2014	95.44	-3.07%
03.01.2014	93.96	-1.56%
06.01.2014	93.43	-0.57%
07.01.2014	93.67	0.26%
08.01.2014	92.33	-1.44%
09.01.2014	91.66	-0.73%
10.01.2014	92.72	1.15%
13.01.2014	91.80	-1.00%

Calculation		
Mean		-0.06%
Variance		0.05%
annualized variance	250	12.70%
monthly variance	12	1.06%
Standard deviation		2.25%
Standard deviation		2.25%
annualized standard deviation	250	35.64%
monthly standard deviation	12	10.29%

Fig. 4 Calculation of the annualized and intra-year standard deviation

Literature and References to the Excel Tool

Ernst, D. (2020): Corporate Risk Management, Course 1—Assignment 9.

Standard literature:

Hopkin, P. (2018): Fundamentals of Risk Management: Understanding, Evaluating and Implementing Effective Risk, 5. Edition, New York.

Hull, J.C. (2018): Risk Management and Financial Institutions, 5. Edition, Hoboken (New Jersey).

See Excel file `Case Study Corporate Risk Management Part 1`, Excel worksheet `Variance & Standard Deviation`

Assignment 10: Calculation of Semi-variance and Semi-standard Deviation

Assignment

Calculate the semi-variance and the semi-standard deviation for the discrete daily oil price returns based on the assumption of a sample.

Content

In contrast to overall risk measures, the *semi-variance* and the *semi-standard deviation* are concentrated in the upper or lower part of a distribution. Semi-variance and semi-standard deviation are therefore one-sided risk measures.

Since both the variance and the standard deviation represent two-dimensional and/or symmetrical risk measures by taking into account positive and negative deviations from a mean value, they do not necessarily correspond to the interest of risk managers. When assessing risks with semi-variance and semi-standard deviation, for example, in the case of commodities that are purchased, the attention is focused solely on the positive deviations from an observed mean. If on the other hand, investments have already been made, e.g., in shares or in oil, the risk lies in the negative deviations from the expected value. In the present example, we consider the negative deviations from the expected value, i.e., the risk in the case of an investment in oil that has already been made.

Important Formulas

The formula for the semi-variance based on a sample is:

$$\text{SemiVar}[r] = \frac{1}{m-1}\sum_{t=1}^{m}\left(r_t^{<m} - \mu\right)^2 \tag{8}$$

SemiVar $[r]$ = Semi-variance of returns
m = Number of observations
$r_t^{<m}$ = Returns that lead to a negative deviation from the mean
μ = Expected rate of return

Workbook: `Case Study Risk Management Part 1` ➲ Worksheet: `Semi-VAR & Semi-STD`

Calculation of the semi-variance:

Excel example: `K9=SUM(H9:H1310)/(COUNT(H9:H1310)-1)`

Analogous to the standard deviation, the semi-standard deviation is obtained according to:

$$\text{SemiS} = \sqrt{\frac{1}{\text{m}-1}\sum_{\text{t}=1}^{\text{m}}\left(r_t^{<m} - \mu\right)^2} \tag{9}$$

SemiS$[r]$ = Semi-standard deviation of return

Calculation of the semi-standard deviation:

Excel example: `K10=SQRT(K9)`

Procedure

- First, the continuous daily returns are calculated from the oil prices (column `D`).
- Then the mean value of the discrete daily returns is calculated (cell `K7`).
- The deviations of the continuous daily returns from the mean value are then calculated (column `F`). Only the negative deviations should be selected here. This is done in Excel by an `IF` function that deletes positive values with the `""` command and leaves negative values `F9=IF(D9-$K$7>0;"";D9-$K$7)`.
- After the deviations are calculated, the negative values are squared `H9=IF(F9="";"";F9^2)`.
- Finally, the semi-variance of the sample `K9=SUM(H9:H1310)/(COUNT(H9:H1310)-1)` and the semi-standard deviation of the sample `K10=SQRT(K9)` are calculated.
- The semi-standard deviation is 2.36%. The risk value with a one-sided consideration of the negative deviations from the expected value is thus somewhat higher than the standard deviation, which shows a value of 2.25%.

Result (Fig. 5)

Asset	Crude Oil
Unit	%

Date	Crude Oil price	Continuous return	$(r_i^{<m}-\mu)$ Oil price	$(r_i^{<m}-\mu)^2$ Oil price
01.01.2014	98.42			
02.01.2014	95.44	-3.07%	-3.02%	0.0909%
03.01.2014	93.96	-1.56%	-1.50%	0.0226%
06.01.2014	93.43	-0.57%	-0.51%	0.0026%
07.01.2014	93.67	0.26%		
08.01.2014	92.33	-1.44%	-1.38%	0.0191%
09.01.2014	91.66	-0.73%	-0.67%	0.0045%
10.01.2014	92.72	1.15%		
13.01.2014	91.80	-1.00%	-0.94%	0.0088%
14.01.2014	92.59	0.86%		
15.01.2014	94.17	1.69%		
16.01.2014	93.96	-0.22%	-0.16%	0.0003%
17.01.2014	94.37	0.44%		

	Oil price
Mean	-0.060%
Semi VAR	0.06%
Semi STD	2.36%

Fig. 5 Calculation of the semi-variance and the semi-standard deviation

Literature and References to the Excel Tool

Ernst, D. (2020): Corporate Risk Management, Course 1—Assignment 10.

Standard literature:

Hopkin, P. (2018): Fundamentals of Risk Management: Understanding, Evaluating and Implementing Effective Risk, 5. Edition, New York.

Hull, J.C. (2018): Risk Management and Financial Institutions, 5. Edition, Hoboken (New Jersey).

See Excel file `Case Study Corporate Risk Management Part 1`, Excel worksheet `Semi-VAR & Semi-STD`

Course Unit 3: Models for Calculating Volatility

Assignment 11: Calculation of the Moving Volatility

Assignment
Calculate the moving volatility as the daily moving volatility for the discrete daily returns of the oil price based on the last 30 and 250 trading days.

Content

Estimates of the volatility vary widely depending on the time periods chosen as the base of the data. To improve the accuracy of the estimation of the variance or standard deviation, statistics usually require as much data as possible. At the same time, however, a constant dispersion, i.e., a constant volatility over time, is assumed. Such volatilities, based on long periods of time, indicate the extent to which a market typically moves over the long term. In order to measure the strength of the ongoing variability, it is a common practice to shorten the length of the observation period and calculate the standard deviation on a rolling basis for a fixed time span. This shortened historical calculation method is then shifted further in steps toward the current time period. This results in a time series whose individual elements are calculated as moving averages and is generally referred to as *moving volatility*.

The most commonly used moving volatilities are the 30- and 250-day volatilities. The standard deviation of the last 29 returns (30 trading days) or the last 249 returns (250 trading days) is calculated. It is to be noted here that the 30- and 250-day volatilities are still daily volatilities. The 30- and 250-days merely indicate the analysis period. The decisive factor is the return. If, as in this case, there is a discrete daily return, the volatility calculated is also a daily one, regardless of how many values are included in the calculation.

(continued)

D. Ernst, J. Häcker, *Corporate Risk Management*, Springer Texts in Business and Economics, https://doi.org/10.1007/978-3-031-53126-2_4

In practice, two assumptions have been established for a simplified calculation of the moving volatility, which we will refer here for the following calculations:

1. The arithmetic mean is set/reset to zero on a daily basis when calculating the moving volatility. This assumption is justified by the fact that the expected change in the market price on a daily basis has no practical relevance.
2. $n-1$ is replaced by n. This change leads from an unbiased estimator to a maximum likelihood estimator, which will play an important role in the following models.

Important Formulas
Formally, moving historical volatility is defined as:

$$\boxed{\sigma_t = \sqrt{\frac{1}{m} \cdot \sum_{i=1}^{m} r_{n-i}^2}} \tag{1}$$

m = Number of observations
n = Period of estimation = $m+1$
$i = 0, 1, 2, 3, \ldots, m$

Workbook: `Case Study Risk Management Part 1` ➲ Worksheet: `Moving Volatility`

Calculation of the 30-day volatility:

Excel example: `F37=SQRT(SUM(E8:E36)/COUNT(E8:E36))`

Calculation of the 250-day volatility:

Excel example: `G257=SQRT(SUM(E8:E256)/COUNT(E8:E256))`

Procedure

- Based on the oil price data, calculate the continuous returns (column `D`) and, taking into account the assumptions made above, the daily variances (column `E`). Then calculate the 30- and 250-day volatility according to the above formula. For the 30-day volatility it is `F37=SQRT(SUM(E8:E36)/NUMBER(E8:E36))` and for the 250-day volatility `G257=SQRT(SUM(E8:E256)/NUMBER(E8:E256))`.
- Please note that the results of the calculations are again daily returns, based on the data for 30 or 250 trading days.

Result (Figs. 1 and 2)

	A	B	C	D	E	F	G
31		04.02.t(1)	97.19	0.79%	0.01%		
32		05.02.t(1)	97.38	0.20%	0.00%		
33		06.02.t(1)	97.84	0.47%	0.00%		
34		07.02.t(1)	99.88	2.06%	0.04%		
35		10.02.t(1)	100.06	0.18%	0.00%		
36		11.02.t(1)	99.94	-0.12%	0.00%		
37		12.02.t(1)	100.37	0.43%	0.00%	1.13%	
38		13.02.t(1)	100.35	-0.02%	0.00%	0.98%	
39		14.02.t(1)	100.30	-0.05%	0.00%	0.94%	
40		17.02.t(1)	100.30	0.00%	0.00%	0.93%	
41		18.02.t(1)	102.43	2.10%	0.04%	0.93%	
42		[illegible]	[illegible]	[illegible]	[illegible]	[illegible]	

Fig. 1 Calculation of the 30-day volatility

	A	B	C	D	E	F	G
251		09.12.t(1)	63.82	1.21%	0.01%	2.75%	
252		10.12.t(1)	60.94	-4.62%	0.21%	2.75%	
253		11.12.t(1)	59.95	-1.64%	0.03%	2.87%	
254		12.12.t(1)	57.81	-3.63%	0.13%	2.89%	
255		15.12.t(1)	55.91	-3.34%	0.11%	2.94%	
256		16.12.t(1)	55.93	0.04%	0.00%	2.98%	
257		17.12.t(1)	56.47	0.96%	0.01%	2.96%	1.46%
258		18.12.t(1)	54.11	-4.27%	0.18%	2.96%	1.45%
259		19.12.t(1)	56.52	4.36%	0.19%	3.06%	1.47%
260		22.12.t(1)	55.26	-2.25%	0.05%	3.15%	1.50%
261		23.12.t(1)	57.12	3.31%	0.11%	3.17%	1.51%
262		24.12.t(1)	55.84	-2.27%	0.05%	3.23%	1.52%

Fig. 2 Calculation of the 250-day volatility

Literature and References to the Excel Tool

Ernst, D. (2020): Corporate Risk Management, Course 1—Assignment 11.

Standard literature:

Hopkin, P. (2018): Fundamentals of Risk Management: Understanding, Evaluating and Implementing Effective Risk, 5. Edition, New York.

Hull, J.C. (2018): Risk Management and Financial Institutions, 5. Edition, Hoboken (New Jersey).

See Excel file Case Study `Corporate Risk Management`, Excel worksheet `Moving Volatility`

Assignment 12: Calculation of the Moving Volatility with Linearly Decreasing Weights and with Exponentially Decreasing Weights

Assignment
Calculate the moving volatility with linearly decreasing weights and exponentially decreasing weights for the discrete daily returns of the oil price based on the last 250 trading days.

Content

When calculating the moving volatility, we have so far weighted the variances—$r_{n-1}^2, r_{n-2}^2, \ldots, r_{n-m}^2$—equally. Since the most recent values are more meaningful when estimating current volatility, it is important to give them a higher weighting. For this purpose, we use linearly decreasing weights and exponentially decreasing weights. This results in *moving volatility with linearly decreasing weights* and *moving volatility with exponentially decreasing weights.*

Important Formulas
The formula for moving volatility with decreasing weights is:

$$\boxed{\sigma_n = \sqrt{\sum_{i=1}^{m} \alpha_i \cdot r_{n-i}^2}} \tag{2}$$

α_i = Weighting factor
m = Number of observations
n = Period of estimation = m+1
$i = 0, 1, 2, 3, \ldots, m$

The variable α_i has a value >0 and <1 and defines the weight given to the observation i days ago. The weights α_i are to be modelled so that recent values receive higher weights and older values lower weights. The sum of all the weights α_i must equal one, i.e.:

$$\boxed{\sum_{i=1}^{m} \alpha_i = 1} \tag{3}$$

The following applies to linearly decreasing weights:

$$\boxed{\alpha_i = \frac{n-i}{\sum_{i=1}^{m} i}} \tag{4}$$

The following applies to exponentially decreasing weights:

$$\boxed{\alpha_i = (1-\lambda) \cdot \lambda^{i-1}} \tag{5}$$

λ = Reduction rate

The weights α_i decrease at the rate λ when we go backward in time. Each weight is λ times the previous weight.

$$\boxed{\alpha_i = \lambda \cdot \alpha_{i+1}} \tag{6}$$

Workbook: `Case Study Risk Management Part 1` ➲ Worksheet: `Vola lin. and exp.dec. weights`

Calculation of the linearly decreasing weights:

Excel example: `L8=J8/SUM($K$8:$K$256)`

Calculation of the weighted 250-day variance:

Excel example: `G257=SUMPRODUCT(E8:E256;$L$8:$L$256)`

Calculation of the 250-day volatility:

Excel example: `H257=SQRT(G257)`

Calculation of exponentially decreasing weights:

Excel example: `R8=(1-$R$3)*$R$3^(Q8-1)`

Calculation of the weighted 250-day variance:

Excel example: `N257= SUMPRODUCT(E8:E256;$R$8:$R$256)`

Calculation of the 250-day volatility:

Excel example: `O257=SQRT(N257)`

Procedure

- Calculate the discrete daily returns in column `D`.
- Calculate the daily variance in column `E` based on the discrete daily returns.
- Calculate in column `L` the weights of the daily variance α_i or linearly decreasing weights `L8=J8/SUM($K$8:$K$256)`.
- Calculate the moving variance with linearly decreasing weights
  `G257=SUMPRODUCT(E8:E256;$L$8:$L$256)`.
- Then calculate the moving volatility with exponentially decreasing weights `H257=SQRT(G257)`.
- Calculate in column `R` the weights of the daily variance α_i using the factor λ in cell `R3` according to the `R8=(1-$R$3)*$R$3^(Q8-1)`.
- Calculate the moving variance with exponentially decreasing weights
  `N257=SUMPRODUCT(E8:E256;$R$8:$R$256)`.
- Then calculate the moving volatility with exponentially decreasing weights `O257=SQRT(N257)`.
- Please note that the results of the calculations are again daily variances and daily standard deviations based on a data basis of 250 trading days.

Result (Figs. 3, 4, 5, and 6)

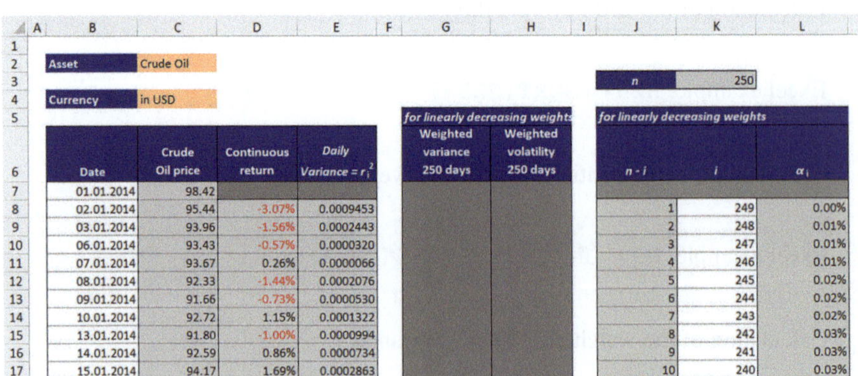

Asset: Crude Oil; Currency: in USD; n: 250

Row	Date	Crude Oil price	Continuous return	Daily Variance = r_i^2	Weighted variance 250 days (for linearly decreasing weights)	Weighted volatility 250 days (for linearly decreasing weights)	$n - i$	i	α_i
7	01.01.2014	98.42							
8	02.01.2014	95.44	-3.07%	0.0009453			1	249	0.00%
9	03.01.2014	93.96	-1.56%	0.0002443			2	248	0.01%
10	06.01.2014	93.43	-0.57%	0.0000320			3	247	0.01%
11	07.01.2014	93.67	0.26%	0.0000066			4	246	0.01%
12	08.01.2014	92.33	-1.44%	0.0002076			5	245	0.02%
13	09.01.2014	91.66	-0.73%	0.0000530			6	244	0.02%
14	10.01.2014	92.72	1.15%	0.0001322			7	243	0.02%
15	13.01.2014	91.80	-1.00%	0.0000994			8	242	0.03%
16	14.01.2014	92.59	0.86%	0.0000734			9	241	0.03%
17	15.01.2014	94.17	1.69%	0.0002863			10	240	0.03%
18	16.01.2014	93.96	-0.22%	0.0000050			11	239	0.04%

Fig. 3 Calculation of the linearly decreasing weights

	A	B	C	D	E	F	G	H	I	J	K	L
250		08.12.2014	63.05	-4.33%	0.0018748					243	7	0.78%
251		09.12.2014	63.82	1.21%	0.0001473					244	6	0.78%
252		10.12.2014	60.94	-4.62%	0.0021323					245	5	0.79%
253		11.12.2014	59.95	-1.64%	0.0002683					246	4	0.79%
254		12.12.2014	57.81	-3.63%	0.0013213					247	3	0.79%
255		15.12.2014	55.91	-3.34%	0.0011168					248	2	0.80%
256		16.12.2014	55.93	0.04%	0.0000001					249	1	0.80%
257		17.12.2014	56.47	0.96%	0.0000923		0.00031055	0.0176224				
258		18.12.2014	54.11	-4.27%	0.0018225		0.000309574	0.0175947				100.00%
259		19.12.2014	56.52	4.36%	0.0018988		0.000322467	0.0179574				
260		22.12.2014	55.26	-2.25%	0.0005083		0.00033592	0.0183281				
261		23.12.2014	57.12	3.31%	0.0010959		0.000338189	0.0183899				
262		24.12.2014	55.84	-2.27%	0.0005137		0.000345143	0.0185780				
263		25.12.2014	55.84	0.00%	0.0000000		0.00034741	0.0186389				

Fig. 4 Calculation of the 250-day volatility for linearly decreasing weights

	M	N	O	P	Q	R
1						
2						
3					Lambda λ	0.95
4						
5		for exponentially decreasing weigh			for exponentially decreasing weights	
6		Weighted variance 250 days	Weighted volatility 250 days		i	α_i
7						
8					249	0.00%
9					248	0.00%
10					247	0.00%
11					246	0.00%
12					245	0.00%
13					244	0.00%
14					243	0.00%
15					242	0.00%
16					241	0.00%

Fig. 5 Calculation of the exponentially decreasing weights

	M	N	O	P	Q	R
250					7	3.68%
251					6	3.87%
252					5	4.07%
253					4	4.29%
254					3	4.51%
255					2	4.75%
256					1	5.00%
257		0.000809342	0.0284489			
258		0.000773491	0.0278117			100.00%
259		0.00082594	0.0287392			
260		0.000879585	0.0296578			

Fig. 6 Calculation of 250-day volatility for exponentially decreasing weights

Literature and References to the Excel Tool

Ernst, D. (2020): Corporate Risk Management, Course 1—Assignment 12.

Standard literature:

Hopkin, P. (2018): Fundamentals of Risk Management: Understanding, Evaluating and Implementing Effective Risk, 5. Edition, New York.

Hull, J.C. (2018): Risk Management and Financial Institutions, 5. Edition, Hoboken (New Jersey).

See Excel file Case Study `Corporate Risk Management`, Excel worksheet `G1. Vola lin. and exp. dec. weights.`

Assignment 13: Volatility Calculation with the EWMA Model

Assignment

Calculate the volatility with the EWMA model for the discrete daily returns of the oil price.

Content

The approach of moving volatility with exponentially decreasing weights has the disadvantage that the weighting factor λ is determined by the financial engineer without a direct reference to the available historical data. This weakness is eliminated by the exponentially weighted moving average model (EWMA model). In the EWMA model, historical values are also exponentially weighted, so that values in the recent past receive a higher weighting than older values. For this purpose, the weighting factor λ is used again, which is also called delay factor. Its value is always between 0 and 1.

However, the EWMA model now estimates the value of the weighting factor λ that best explains the historical data of the conditional EWMA variances. For this, we use the *maximum likelihood method*. The maximum likelihood method calculates the value of the parameter λ, which maximizes the probability of occurrence of the historical values (here EWMA variance). The parameter λ is determined so that it best describes the observations made so far.

Important Formulas

The formula for the estimated value of the variance according to the EWMA model is:

$$\boxed{\sigma_n^2 = \lambda \cdot \sigma_{n-1}^2 + (1-\lambda) \cdot r_{n-1}^2} \tag{7}$$

λ = Weighting factor

The EWMA model shows that with only two values—the estimated variance for the previous day σ_{n-1}^2 and the actual rate of change on the previous day r_{n-1}^2—the variance for the next day can be estimated. The estimated value for the variance of the previous day includes historical values of daily changes, which are taken into account with exponentially decreasing weights that decrease the further they are in the past. For the demonstration of this fact, the resolution of the recursion formulas is suitable.

$$\boxed{\sigma_n^2 = (1-\lambda) \cdot \sum \left(\lambda^{i-1} \cdot r_{n-i}^2 + \lambda^n \cdot \sigma_0^2\right) \text{ with } i = 1 \text{ to } n} \tag{8}$$

The sum product of $\lambda^n \cdot \sigma_0^2$ approaches zero, because λ is regularly less than one and λ^n approaches zero with increasing n. This leads to the following simplification:

$$\boxed{\sigma_n^2 = (1-\lambda) \cdot \sum \left(\lambda^{i-1} \cdot r_{n-i}^2\right) \text{ with } i = 1 \text{ to } n} \tag{9}$$

The weighting factor λ^{i-1} for the historical rates of change r_{n-i}^2 becomes smaller the larger i is, i.e., the further the observation lies in the past.

We now consider how the maximum likelihood method can be used to estimate the parameter λ. r_i^2 is the observed variance and σ_i^2 is the estimated variance. We assume that the expected value of the observed variances is zero and their probability distribution is a normal distribution.

The probability that the observations occur in the order of time in which they were observed is calculated with the following formula:

$$\boxed{\prod_{i=1}^{m} \left[\frac{1}{\sqrt{2\pi\sigma_i^2}} \exp\left(\frac{-r_i^2}{2\sigma_i^2} \right) \right]} \tag{10}$$

In the maximum likelihood method, the value of σ_i^2, which maximizes the above expression, is the best estimator. By using logarithms, we can simplify the above formula as follows:

$$\boxed{\sum_{i=1}^{m}\left(-\ln(\sigma_i^2)-\frac{r_i^2}{\sigma_i^2}\right)} \tag{11}$$

With the iterative search method (SOLVER in Excel) we can now determine the parameter λ, which maximizes the last listed formula.

Workbook: Case Study Risk Management PHARMA GROUP ➲ Worksheet: EWMA

Calculation of the daily variance:

Excel example: E12=D12^2

Calculation of the EWMA variance (start value):

Excel example: F12=E12

Calculation of the EWMA variance (subsequent values):

Excel example: F13=F12*$D$7+(1-$D$7)*E12

Calculation of the EWMA volatility:

Excel example: I12=SQRT(F12)

Calculation of the probability that the estimated value will occur:

Excel example: G12=-LN(F12)-E12/F12

Sum of the probabilities, which is maximized:

Excel example: G7=SUM(G12:G1314)

Output value for λ, which is used in the SOLVER for optimization:

Excel example: C7='Assumptions General'!C176

Procedure

- Calculate the discrete daily returns in column `D`.
- Calculate the daily variance in column `E` based on the discrete daily returns.
- In the next step, the EWMA variance is calculated. For this, we need the value for λ, which we first obtain from the worksheet `Assumptions General`. It is linked to cell `C7`. We link this cell in turn to cell `D7`, which after optimization with the solver will contain the value λ resulting from the optimization.
- For the calculation of the EWMA variance we need an initial value. Here, we insert the variance from cell `E12` to cell `F12`.
- In `F13`, the actual formula for the EWMA variance is then applied `F13=F12*$D$7+(1-$D$7)*E12`.
- The EWMA volatility is the square root of the EWMA variance `I12=SQRT(F12)`.
- Column `G` contains the probability calculation for the maximum method. The Excel formula is `G12=-LN(F12)-E12/F12`.
- In cell `G7`, the probabilities are summed up `G7=SUM(G12:G1314)`, which are then maximized during optimization.
- For optimization using the `SOLVER`, the following values must be entered in the `SOLVER` (Fig. 7).

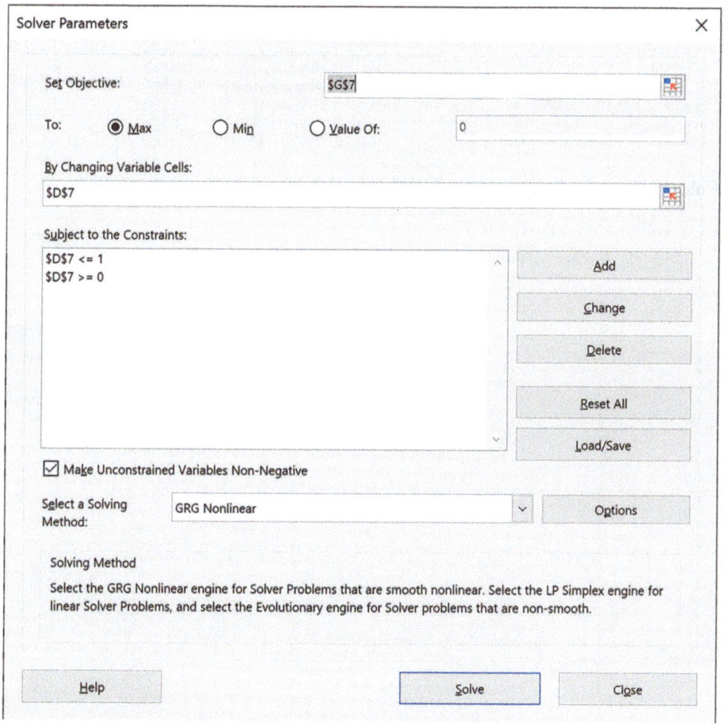

Fig. 7 Solver parameter for the EWMA model

- This results in a value for λ of 0.939251. This λ best describes the historical observations of the EWMA variance.
- If this value is used as λ when calculating the moving volatility with exponentially decreasing weights, the same values for the weighted variance are obtained as for the EWMA variance.

Result (Fig. 8)

	B	C	D	E	F	G	H	I
2	Asset	Crude Oil						
4	Currency	in USD						
6		Starting values	Optim. values					
7	λ	0.95000	0.939251			8872.464		

	Date	Crude Oil price	Continuous return	Daily Variance = r_i^2	EWMA-Variance	Likelihood		EWMA-Volatility
11	01.01.2014	98.42						
12	02.01.2014	95.44	-3.07%	0.09%	0.000945332	5.964		0.030746258
13	03.01.2014	93.96	-1.56%	0.02%	0.000945332	6.706		0.030746258
14	06.01.2014	93.43	-0.57%	0.00%	0.000902743	6.975		0.030045678
15	07.01.2014	93.67	0.26%	0.00%	0.000849846	7.063		0.029152121
16	08.01.2014	92.33	-1.44%	0.02%	0.000798619	6.873		0.028259846
17	09.01.2014	91.66	-0.73%	0.01%	0.000762716	7.109		0.027617317
18	10.01.2014	92.72	1.15%	0.01%	0.000719604	7.053		0.026825444
19	13.01.2014	91.80	-1.00%	0.01%	0.000683921	7.142		0.02615188
20	14.01.2014	92.59	0.86%	0.01%	0.000648414	7.228		0.02546398
21	15.01.2014	94.17	1.69%	0.03%	0.000613484	6.930		0.024768618
22	16.01.2014	93.96	-0.22%	0.00%	0.000593609	7.421		0.024364085
23	17.01.2014	94.37	0.44%	0.00%	0.00055785	7.457		0.023618859

Fig. 8 Calculating volatility using the EWMA model

Literature and References to the Excel Tool

Ernst, D. (2020): Corporate Risk Management, Course 1—Assignment 13.

Standard literature:

Hopkin, P. (2018): Fundamentals of Risk Management: Understanding, Evaluating and Implementing Effective Risk, 5. Edition, New York.

Hull, J.C. (2018): Risk Management and Financial Institutions, 5. Edition, Hoboken (New Jersey).

See Excel file Case Study `Corporate Risk Management`, Excel worksheet `EWMA`.

Assignment 14: Volatility Calculation with the ARCH Model

Assignment

Calculate the volatility with the ARCH model for the discrete daily returns of the oil price.

Content

Most theoretical approaches in capital market theory assume that the variance of returns is constant over time. However, a closer look at the rates of price change in speculative markets reveals that although they fluctuate around a constant average, their variability over time is not constant. There is rather a tendency towards volatility clustering. Volatility clustering describes the temporal concentration of absolutely high and absolutely low returns. This means that the development of volatility over time follows a certain pattern. A phase of high volatility is followed by a phase of low volatility and vice versa. Volatility clusters therefore form in each phase. Classical methods such as regression analysis or time series analysis assume a constant variance of the forecast errors over time and are unsuitable to explain phenomena such as volatility clustering. Especially, stock and exchange rates or interest rates show such behavioral patterns, which require a non-linear modelling approach. Such an approach was developed by *Engle*, who received the Nobel Prize in Economics in 2003 for his work on ARCH models.

The abbreviation ARCH stands for "Autoregressive Conditional Heteroskedasticity." ARCH models also try to take into account that volatilities follow a certain pattern. The property of changing volatilities over time is

(continued)

called heteroskedasticity. ARCH models are also autoregressive, i.e., volatility is measured in relation to its predecessor, i.e., it is conditional.

The ARCH(*m)* model considered here can be seen as an extension of the method of moving volatility that we have already discussed. The extension is that in the calculation of the estimated variance, in addition to the weighted historical variance, there is also a long-term variance to which the estimated variance has a long-term impact. Here too, the maximum likelihood method is used to determine the parameters necessary for the model.

ARCH models are particularly well-suited for short-term forecasts based on current historical data. Here, we consider the data of the past 5 days for the ARCH model. Methodologically, it would make sense to use continuous returns here. However, in order to stick to the delimitation of the areas of application of discrete and continuous returns we made at the beginning of the case study and to enable a comparability of the approaches for calculating volatilities, we use discrete returns as the basis of our calculations in this ARCH model as well as in the following GARCH model. However, discrete returns can easily be replaced by continuous returns.

Important Formulas
The formula for the variance σ^2 at time n in the ARCH(m) model is:

$$\boxed{\sigma_n^2 = \gamma \cdot V_L + \sum_{i=1}^{m} \alpha_i \cdot r_{n-1}^2} \tag{12}$$

V_L = Average, long-term variance of the time series
γ = Weighting factor of V_L
r_{n-i}^2 = The squared return actually measured on the previous day (= variance)
α_i = Weighting factor on day i
m = Number of observations
n = Period of estimation = m+1

The ARCH(m) model implies that the variance σ_t^2 can be explained by the average long-term variance and m observations.

Since the sum of the weights is one, the following applies:

$$\boxed{\gamma + \sum_{i=1}^{m} \alpha_i = 1} \tag{13}$$

In the ARCH(m) model, most recent observations are assigned a higher weight and older observations a lower weight. With $\omega = \gamma \cdot V_L$, the above formula can also be written as follows:

$$\boxed{\sigma_n^2 = \omega + \sum_{i=1}^{m} \alpha_i \cdot r_{n-1}^2} \quad (14)$$

Workbook: `Case Study Risk Management PHARMA GROUP` ➲ Worksheet: `ARCH`

Calculation of the daily variance:

Excel example: `E12=D12^2`

Starting value for ω, which is used in `SOLVER` for optimization:

Excel example: `L3='Assumptions General'!C181`

Calculation of the weights $\alpha(1)$:

Excel example: `L4='Assumptions General'!C182`

Calculation of the conditional variance:

Excel example: `F17=$M$3+$M$4*E12+$M$5*E13+$M$6*E14+$M$7*E15+$M$8*E16`

Calculation of the conditional volatility:

Excel example: `I17=SQRT(F17)`

Calculation of the probability of the estimated value occurring:

Excel example: `G17=-LN(F17)-E17/F17`

Sum of the probabilities, which is maximized:

Excel example: `G8=SUM(G12:G1314)`

Calculation of the weighting factor of γ_L:

Excel example: `I2=1-SUM(M4:M8)`

Calculation of the daily, long-term variance V_L:

Excel example: `I3=M3/I2`

Calculation of the daily, long-term volatility:

Excel example: `I4=SQRT(I3)`

Calculation of the annual, long-term volatility:

Excel example: `I5=I4*SQRT('Assumptions General'!C187)`

Procedure

- Calculate the discrete daily returns in column `D`.
- Calculate the daily variance in column `E` based on the discrete daily returns.
- In the next step, the conditional variance is calculated. For this, we need starting values for the variables ω and α_i, which we obtain from `Assumptions General`. They are linked to the cells `L3` and `L4:L8`. We link these cells again with the cells `M3` and `M4:M8`, which will contain the values from λ and α_i after optimization with the solver.
- In `F17`, the actual formula for the calculation of the conditional variance is executed, which consists of the average, long-term variance and m observations `F17=$M$3+$M$4*E12+$M$5*E13+$M$6*E14+$M$7*E15+$M$8*E16`.
- The conditional volatility is the square root of the conditional variance `I17=SQRT(F17)`.
- Column `G` contains the probability calculation for the maximum likelihood method. The Excel formula is `G17=-LN(F17)-E17/F17`.
- In cell `G8`, the probabilities are summed `G8=SUM(G12:G1314)`, which are then maximized during optimization.
- For optimization using the `SOLVER`, the following values must be entered in the `SOLVER` (Fig. 9).

- This gives a value for ω as 0.000139 and values for α_1 as 0.143471, α_2 as 0.107560, α_3 as 0.297855, α_4 as 0.116534, and α_5 as 0.180002. These values best describe the historical observations of the conditional variance.

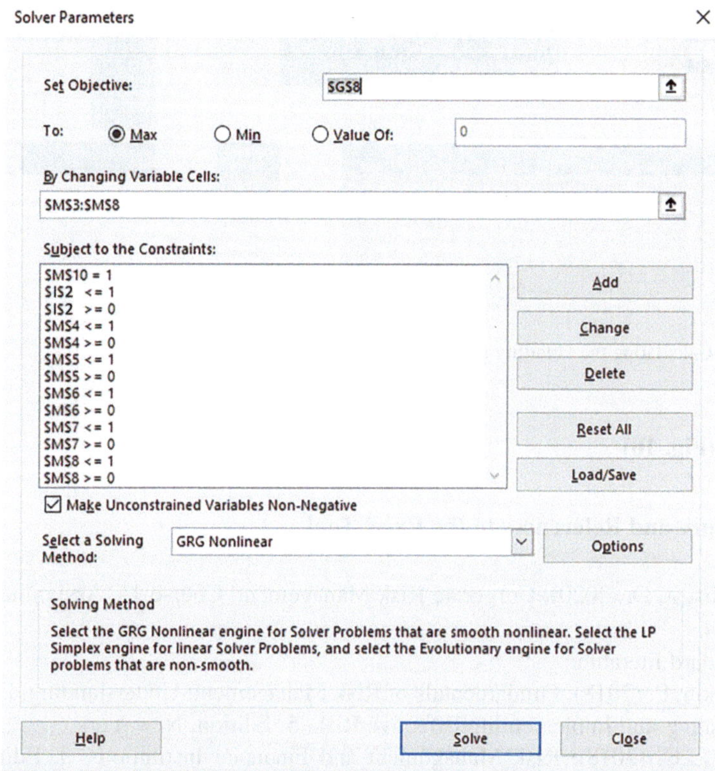

Fig. 9 Solver parameter for the ARCH model

- In the next step, the weighting factor of γ_L `I2=1-SUM(M4:M8)`, the daily long-term variance V_L `I3=M3/I2`, and the daily long-term volatility `I4=SQRT(I3)` can now be calculated.
- The result of the ARCH model used here exhibits that the long-term volatility per year is 47.34%. The weights from the last 5 days of actual measured variances contribute 84.54% while the average long-term variance of the time series contributes 15.46%.

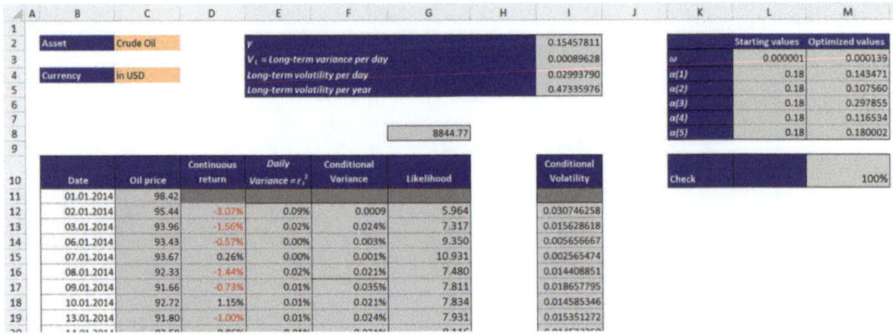

Asset	Crude Oil
Currency	in USD

γ	0.15457811
V_L = Long-term variance per day	0.00089628
Long-term volatility per day	0.02993790
Long-term volatility per year	0.47335976

	Starting values	Optimized values
ω	0.000001	0.000139
α(1)	0.18	0.143471
α(2)	0.18	0.107560
α(3)	0.18	0.297855
α(4)	0.18	0.116534
α(5)	0.18	0.180002
Check		100%

Date	Oil price	Continuous return	Daily Variance = r_i^2	Conditional Variance	Likelihood	Conditional Volatility
					8844.77	
01.01.2014	98.42					
02.01.2014	95.44	-3.07%	0.09%	0.0009	5.964	0.030746258
03.01.2014	93.96	-1.56%	0.02%	0.024%	7.317	0.015628618
06.01.2014	93.43	-0.57%	0.00%	0.003%	9.350	0.005656667
07.01.2014	93.67	0.26%	0.00%	0.001%	10.931	0.002565474
08.01.2014	92.33	-1.44%	0.02%	0.021%	7.480	0.014408851
09.01.2014	91.66	-0.73%	0.01%	0.035%	7.811	0.018657795
10.01.2014	92.72	1.15%	0.01%	0.021%	7.834	0.014585346
13.01.2014	91.80	-1.00%	0.01%	0.024%	7.931	0.015351272

Fig. 10 Calculating the volatility using the ARCH model

Result (Fig. 10)

Literature and References to the Excel Tool

Ernst, D. (2020): Corporate Risk Management, Course 1—Assignment 14.

Standard literature:

Hopkin, P. (2018): Fundamentals of Risk Management: Understanding, Evaluating and Implementing Effective Risk, 5. Edition, New York.

Hull, J.C. (2018): Risk Management and Financial Institutions, 5. Edition, Hoboken (New Jersey).

See Excel file Case Study `Corporate Risk Management`, Excel worksheet `ARCH`.

Assignment 15: Volatility Calculation with the GARCH Model

Assignment

Calculate the volatility with the GARCH model for the discrete daily returns of the oil price.

Content

The idea of the ARCH model has been further developed in various ways and is now one of the most advanced methods in econometrics. *GARCH models* are a generalization (generalized autoregressive conditional heteroscedasticity),

(continued)

developed by *Bollerslev* in 1986. Unlike the ARCH model, the conditional variance of GARCH models depends not only on the history of the time series, but also on its own past. GARCH models also allow volatility clusters to be considered when forecasting volatilities. This exhibits the property of volatilities to follow a certain pattern over time, which is called heteroskedasticity. Furthermore, autoregressive properties can be observed in volatility clusters, i.e., volatility is in turn dependent on its predecessor. If an autoregressive time series is present, the variance is conditional, since its magnitude depends on the value of the previous variance. Simple GARCH models also assume that the underlying rates of change (returns) are normally distributed (there are also GARCH models that do not assume a normal distribution of returns, e.g., EGARCH and GJR). However, their variance can fluctuate over time and depends on the volatility of previous periods. In this way, both the clustering of volatilities and leptokurtic distributions can be modelled. On the one hand, leptokurtic distributions have fat tails, which in contrast to the normal distribution make extreme price changes more probable, and on the other hand they have more values around the expected value, i.e., the peak of the participation is higher and narrower (thin waists) than in a normal distribution.

The estimation of parameters γ, α, and β is mentioned in the literature as a constraint in the use of GARCH models. The parameters are estimated using the maximum likelihood method. The parameters γ, α, and β are estimated on the basis of historical values and in such a way that they maximize the probability of the occurrence of the historical values. The parameters are, therefore, determined in such a way that they best describe the observations made so far.

A GARCH(1,1) model is a special model for analysing and predicting the volatility of time series, particularly in the financial industry. The numbers (1,1) refer to the orders of the model. The first number (1) is the ARCH term. It represents the influence of past error squares on the current variance.. The second number (1) is the GARCH term. It represents the influence of past variance on the current variance. If the parameters are chosen appropriately, the GARCH(1,1) can be transferred to the simpler EWMA model. For the case $\gamma = 0$, $\alpha = 1 - \lambda$ and $\beta = \lambda$, the GARCH model results in the EWMA model, so that the latter can be regarded as a special case of GARCH(1,1).

Important Formulas

The formula for the variance σ^2 at time n in the GARCH(1,1) model is:

$$\boxed{\sigma_n^2 = \gamma \cdot V_L + \alpha \cdot r_{n-1}^2 + \beta \cdot \sigma_{n-1}^2} \tag{15}$$

V_L = Average, long-term variance of the time series
γ = Weighting factor of V_L
r_{t-1}^2 = The squared return actually measured on the previous day (= variance)
α = Weighting factor of r_{t-1}^2

σ_{t-1}^2 = Estimated value for the variance of the previous day
β = Weighting factor of σ_{t-1}^2

The sum of the three weighting factors must equal to one ($\gamma + \alpha + \beta = 1$).

Workbook: `Case Study Risk Management PHARMA GROUP` ➲ Worksheet: `GARCH`

Calculation of the daily variance:

Excel example: `E12=D12^2`

Starting value for ω, which is used in `SOLVER` for optimization:

Excel example: `L3='Assumptions General'!C192`

Starting value for α, which is used in `SOLVER` for optimization:

Excel example: `L4='Assumptions General'!C193`

Starting value for β, which is used in `SOLVER` for optimization:

Excel example: `L5='Assumptions General'!C194`

Calculation of the conditional variance (starting value):

Excel example: `F12=E12`

Calculation of the conditional variance (subsequent values):

Excel example: `F13=$M$3+$M$4*E12+$M$5*F12`

Calculation of the conditional volatility:

Excel example: `I12=SQRT(F12)`

Calculation of the probability of the occurrence of the estimated value:

Excel example: `G12=-LN(F12)-E12/F12`

Sum of the probabilities, which is maximized:

Excel example: `G8=SUM(G12:G1314)`

Calculation of the weighting factor of Υ:

Excel example: `I2=1-M4-M5`

Calculation of the daily, long-term variance V_L:

Excel example: `I3=M3/I2`

Calculation of the daily, long-term volatility:

Excel example: `I4=SQRT(I3)`

Calculation of the annual, long-term volatility:

Excel example: `I5=I4*SQRT('Assumptions General'!C195)`

Procedure

- Calculate the discrete daily returns in column `D`.
- Calculate the daily variance in column `E` based on the discrete daily returns.
- In the next step, the conditional variance is calculated. For this, we need the variables for ω, α and β starting values, which we first obtain from the `Assumptions General`. They are linked to the cells `L3`, `L4`, and `L5`. We link this cell again with the cells `M3`, `M4`, and `M5`, which will contain the values from ω, α and β after optimization with the solver.
- For the calculation of the conditional variance, we need a starting value. Here, we insert the variance from the cell `E12` to the cell `F12`.

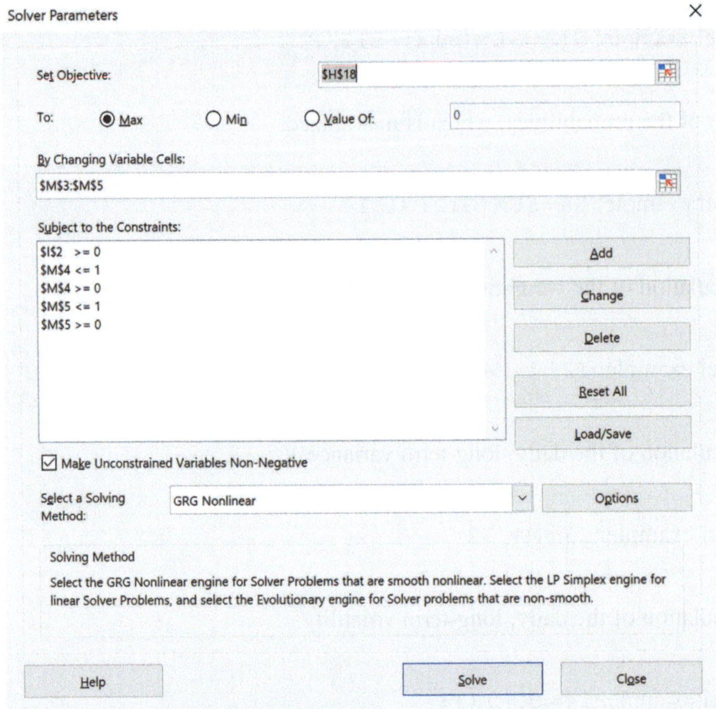

Fig. 11 Solver parameter for the GARCH model

- In `F13`, the actual calculation of the conditional variance then takes place, which depends not only on the variance of the time series, but also on its own past, i.e., the conditional variance of the previous period `F13=$M$3+$M$4*E12+$M$5*F12`.
- The conditional volatility is the square root of the conditional variance `I12=SQRT(F12)`.
- Column `G` contains the probability calculation for the maximum likelihood method. The Excel formula is `G12=-LN(F12)-E12/F12`.
- In the cell `G8`, the probabilities are summed up `G8=SUM(G12:G1314)`, which are then maximized during optimization.
- For optimization using the `SOLVER`, the following values must be entered in the `SOLVER` (Fig. 11):

- This results in values for ω of 0.000005, for α of 0.084084 and for β of 0.912088.
- In the next step the weighting factor of V_L `I2=1-M4-M5`, the daily long-term variance V_L `I3=M3/I2` and the daily long-term volatility `I4=SQRT(I3)` can be calculated. The long-term volatility per year is 54.64%.

The result of the GARCH model used here exhibits that the long-term volatility per year is 54.64%. The conditional variance of the previous day contributes 8.41%, the variance of the previous day 91.21% and the average long-term variance of the time series 0.38% to this result.

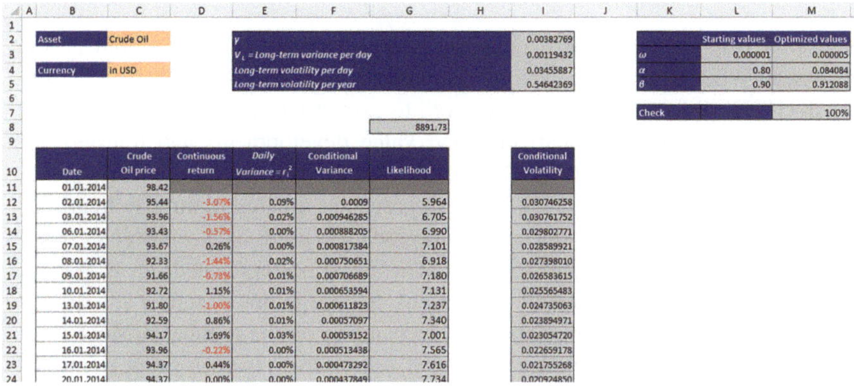

Asset	Crude Oil
Currency	in USD

γ	0.00382769
V_L = Long-term variance per day	0.00119432
Long-term volatility per day	0.03455887
Long-term volatility per year	0.54642369

	Starting values	Optimized values
ω	0.000001	0.000005
α	0.80	0.084084
β	0.90	0.912088
Check		100%

Date	Crude Oil price	Continuous return	Daily Variance = r_i^2	Conditional Variance	Likelihood	Conditional Volatility
					8891.73	
01.01.2014	98.42					
02.01.2014	95.44	-3.07%	0.09%	0.0009	5.964	0.030746258
03.01.2014	93.96	-1.56%	0.02%	0.000946285	6.705	0.030761752
06.01.2014	93.43	-0.57%	0.00%	0.000888205	6.990	0.029802771
07.01.2014	93.67	0.26%	0.00%	0.000817384	7.101	0.028589921
08.01.2014	92.33	-1.44%	0.02%	0.000750651	6.918	0.027398010
09.01.2014	91.66	-0.73%	0.01%	0.000706689	7.180	0.026583615
10.01.2014	92.72	1.15%	0.01%	0.000653594	7.131	0.025565483
13.01.2014	91.80	-1.00%	0.01%	0.000611823	7.237	0.024735063
14.01.2014	92.59	0.86%	0.01%	0.00057097	7.340	0.023894971
15.01.2014	94.17	1.69%	0.03%	0.00053152	7.001	0.023054720
16.01.2014	93.96	-0.22%	0.00%	0.000513438	7.565	0.022659178
17.01.2014	94.37	0.44%	0.00%	0.000473292	7.616	0.021755268
20.01.2014	94.37	0.00%	0.00%	0.000437849	7.734	0.020924850

Fig. 12 Calculation of volatility using the GARCH model

Result (Fig. 12)

Literature and References to the Excel Tool

Ernst, D. (2020): Corporate Risk Management, Course 1—Assignment 15.

Standard literature:

Hopkin, P. (2018): Fundamentals of Risk Management: Understanding, Evaluating and Implementing Effective Risk, 5. Edition, New York.

Hull, J.C. (2018): Risk Management and Financial Institutions, 5. Edition, Hoboken (New Jersey).

See Excel file Case Study `Corporate Risk Management`, Excel worksheet `GARCH`.

Assignment 16: Forecasting Value and Price Developments Using Stochastic Processes

Task

Simulate the oil prices for the given data series for 1 year using an appropriate stochastic process.

GARCH models are used to model short-term price and value developments of assets. When forecasting the long-term prices and the value developments, the stochastic processes are structured well to extract the special features and the characteristics of historical price and value developments and transferring them to the future. A *stochastic process* (also called a random process) is the mathematical representation of random processes ordered in time. In financial market models, special stochastic processes are used, each of which corresponds to an underlying distribution. These can be segregated, for example, according to whether they are continuous, exhibit jumps or reflect a return to the mean (mean-reverting behavior). Depending on the number of stochastic influencing variables included in a model, one refers to them as single or multi-factor models.

In order to simulate the oil price and the interest rate in the context of the P&L calculation, we need models that allow us to extrapolate the price trends and fit well with the historical data and the general characteristics of the price trends. We consider a forecast horizon of 1 year.

From the multitude of stochastic processes, we select five processes that are of prime importance in risk management practice. These processes have different properties that make them suitable for certain applications. According to these criteria, the appropriate process can then be selected.

Content

Table 1 shows five stochastic process types with their properties

Table 1 Process types and their properties

Name	Sign	Mean reversion	Long-term slope behavior of the quantiles
Wiener process	Positive and negative possible	No	Square root plus linear
Brownian motion	Positive and negative possible	No	Square root
Geometric Brownian motion	Only positive possible	No	Exponential
Black-Karasinski process	Only positive possible	Yes	Stationary
Vasicek process	Positive and negative possible	Yes	Stationary

Important Formulas

Wiener Process

A Wiener process is a continuous-time *stochastic process* that has *normally distributed* independent increments.

$$\boxed{dW_t = \varepsilon \cdot \sqrt{dt}} \tag{16}$$

W_t = Wiener process
d = Differential
ε = Random variable, which is standard normally distributed
t = Course of time
dt = Small change in time

The Wiener process is characterized by the following properties:

- *Property 1*: $W_0 = 0$
- *Property 2*: The Wiener process $(W_t)_{t \geq 0}$ has independent increments, i.e., for the points in time $t_1 < t_2 < t_3 < \ldots < t_n$ the increments are $W_{t_n} - W_{t_{n-1}}; W_{t_{n-1}} - W_{t_{n-2}}; \ldots; W_{t_2} - W_{t_1}$ stochastically independent.
- *Property 3*: For all $0 \leq s < t$, the following holds $W_t - W_s \sim N(0, t - s)$. The increments are normally distributed with the expected value $E\,[W_t - W_s] = 0$ and the variance $Var[W_t - W_s] = t - s$.
- *Property 4*: All (single) paths of $(W_t)_{t \geq 0}$ are continuous.
- For small time intervals dt, the increases are dW independent.

The Wiener process is a process with the *Markov property*. This states that, for the prediction of future realizations of the variables W, only its present value is relevant. The variable thus has "no memory," so that the past realizations have no influence on the future developments and are also not suitable for forecasting the future developments.

Brownian Motion

The Brownian motion is a Wiener process, which additionally has a drift μ and a standard deviation σ is exhibited.

The performance of a Brownian motion can be represented by the following formula:

$$\boxed{X_t = \mu \cdot t + \sigma \cdot W_t} \tag{17}$$

X_t = Random variable X at the time t
μ = Drift
t = Course of time
σ = Volatility
W_t = Wiener process

Brownian motion is the solution of (represented by) the following differential equation:

$$\boxed{dX_t = \mu \cdot dt + \sigma \cdot dW_t} \tag{18}$$

Geometric Brownian Motion

Geometric Brownian motion is a continuous stochastic process based on the Brownian motion. The Geometric Brownian motion is popular in modelling financial assets such as stocks, and unlike Brownian motion it always has positive values. Therefore, the random variable X is often replaced by the variable S, which is derived from the English term "stock." The Geometric Brownian motion can be used to model independent, multiplicative increases.

$$\boxed{S_t = S_0 \cdot \exp\left[\left(\mu - \frac{\sigma^2}{2}\right) \cdot t + \sigma \cdot dW_t\right]} \tag{19}$$

S_t = Value at time t
S_0 = Start value at time 0
μ = Drift
σ = Volatility
W_t = Wiener process
t = Course of time

Geometric Brownian motion is the solution of the following stochastic differential equation. The solution is derived with the help of the Itô's Lemma.

$$\boxed{dS_t = S_t \cdot (\mu \cdot dt + \sigma \cdot dW_t)} \tag{20}$$

S_t = Value at time t
dS_t = Small change of the value
d = Differential
dt = Small change in time

Geometric Brownian motion has the following properties:

- *Property 1*: Percentage changes in the stock price are normally distributed, while the absolute price fluctuations follow a lognormal distribution.
- *Property 2*: Stock prices cannot assume negative values. This follows, formally, from their lognormal distribution. If a price of zero is reached, the process remains at this level.

If the dividend payments are also added to the model, the equation of Geometric Brownian motion changes as follows:

$$\boxed{S_t = S_0 \cdot \exp\left[\left(\mu - \delta - \frac{\sigma^2}{2}\right) \cdot t + \sigma \cdot W_t\right]} \tag{21}$$

S_0 = Start value

$\delta =$ Annualized, steady dividend yield of the underlying asset

Ornstein-Uhlenbeck Process

Another class of stochastic processes is the so-called mean reversion process. In addition to a diffusion parameter, they take into account a drift parameter that leads the process to a long-term mean reversion level (trend level).

The simplest example of a mean reversion process is the Ornstein-Uhlenbeck process:

$$\boxed{dX_t = \alpha \cdot (\mu - X_t) \cdot dt + \sigma \cdot dW_t} \qquad (22)$$

$X_t =$ Random variable X at time t.
$d =$ Differential
$\alpha =$ Mean Reversion Speed $=$ (stiffness)
$\mu =$ Mean Reversion Level $=$ Trend Level
$\sigma =$ Diffusion $=$ Volatility
$W_t =$ Wiener process
$t =$ Course of time
$dt =$ Small change in time

If the diffusion σ equal to 0, then the random perturbation of the attraction to μ would be eliminated and the process would converge exponentially towards the mean reversion level.

Black-Karasinski Process

The Black-Karasinski process is a logarithmic model and therefore always generates positive values. It is therefore suitable for the underlying variables that cannot assume negative values. In contrast to the Geometric Brownian motion, whose quantiles take on an exponential course and thus quickly generate value developments that lie outside realistic orders of magnitude, the Black-Karasinski process has a mean reversion process that establishes a motion in the direction of drift. The Black-Karasinski process is based on a stationary process.

The Black-Karasinski process has the following formula:

$$\boxed{d \ln (X_t) = \alpha \cdot (\mu - \ln (X_t)) \cdot dt + \sigma \cdot dW_t} \qquad (23)$$

$X_t =$ Random variable X at time t
$d =$ Differential
$\alpha =$ Mean Reversion Speed
$\mu =$ Mean Reversion Level $=$ Trend Level (long-term trend $= \exp(\mu)$)
$\sigma =$ Diffusion $=$ Volatility
$W_t =$ Wiener process

Vasicek Process

The Vasicek process is very similar in structure to the Black-Karasinski process. The Vasicek process is not a logarithmic model and therefore, allows this process to be used for the assets with positive and negative values. The Vasicek process also exhibits mean reversion. This means that the random variable is always attracted to the mean reversion level. If the random variable X is above the long-run mean at the time t then the drift term is negative, i.e., there is an attraction from above against the mean reversion level. If it is below, the drift term is, correspondingly, positive and there is an attraction from below against the Mean Reversion Level. The parameter α determines the speed of the return to the Mean Reversion Level. The parameter σ specifies the random influence of the Wiener process.

$$\boxed{dX_t = \alpha \cdot (\mu - X_t) \cdot dt + \sigma \cdot dW_t} \tag{24}$$

X_t = Random variable x at time t
d = Differential
α = Mean Reversion Speed
μ = Mean Reversion Level = trend level (long-term trend = exp(μ)) = "long term mean reversion level".
σ = Diffusion = Volatility
W_t = Wiener process

Workbook: `Case Study Risk Management Part 1` Worksheet: ➲ `Black-Karasinski`

The assignment is carried out using the Excel Add-In Risk Kit.

Black-Karasinski Process

When calibrating the oil prices for the Black-Karasinski process with Risk-Kit, the following supplementary function is applied:

Excel example: `{F6=CalibrateProcess(F4;C3:C1306;1)}`

For the simulation, we use the supplementary function "`BK_path`" for the Black-Karasinski process, which uses the parameters in the same order as returned by the function "`CalibrateProcess`."

Excel-Beispiel: `{G13=BK_path(C3;F7;G7;H7;F13:F25)}`

The results of the Monte Carlo simulation are generated and displayed using the following function:

Excel example: `I15=ProcessDefinition(F13:F25;G13:G25;G12)`

The results of the Monte Carlo simulation can also be displayed graphically. The function for this is:

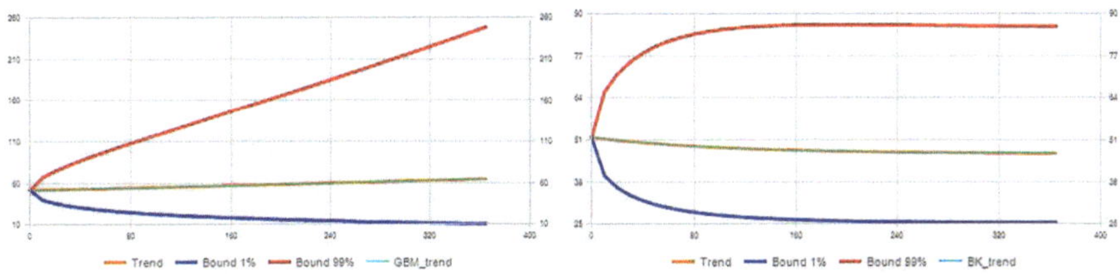

Fig. 13 Calibrated geometric Brownian Motion and Black-Karasinski process

Excel example: `I16=+PlotProcess(I15)`

Procedure

- First, we take the data for the price of oil.
- Using the above set of criteria, we select which stochastic process best fits the characteristics of the data.

  In the case of the oil price, we note that it can only take on positive values. Therefore, the Wiener process, Brownian motion, and the Vasicek process are not useful in this case.

  The remaining processes are the Geometric Brownian Motion and the Black-Karasinski process. Therefore, it is now necessary to examine more closely which of these two processes best fits the price/value behavior of the oil price.
- In the time window considered, the oil price exhibits quite stationary behavior, with no clear long-term trend. Calibrated to the data, the Geometric Brownian motion and the Black-Karasinski process show quite different forecast curves. In particular, the Geometric Brownian Motion shows exponentially increasing quantiles. These quantiles reach oil price levels that are far outside the range of the historical data. The confidence bands of the Black-Karasinski process, on the other hand, cover the historical range of oil prices in a reasonably conservative manner. Therefore, we choose the Black-Karasinski model for the 1-year forecast of the oil price (Fig. 13).

- To model the oil price with the Black-Karasinski process, we first enter the name of the relevant process in cell `F4`. Since the relevant process is the Black-Karasinski process we enter the word `BLACKKARASINSKI`.
- We then calibrate the historical oil price data to obtain the parameters for the Black-Karasinski process.
- It is important that we first completely select the output range of the calibration (cells `F6:H7`) and make sure that these cells remain selected during the input of the Risk-Kit function.
- Then we enter the following Risk-Kit function in the cell window: `=CalibrateProcess(F4;C3:C1306;1)`.

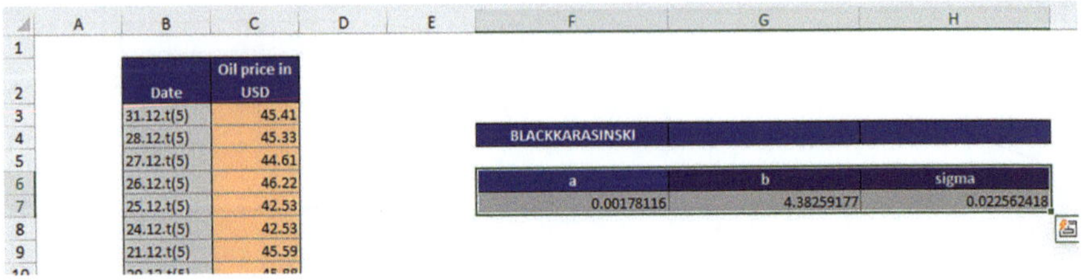

Fig. 14 Calibration of historical data for the Black-Karasinski process

- `DeltaT` must be entered in the Risk-Kit function, as the process parameters are scaled accordingly. Since the oil price is a daily value, `DeltaT` = 1 is set (Fig. 14).
- We end the command with a curly bracket by holding down the `CTRL+ALT` keys and then pressing `ENTER`. In the cell window the following will appear: {`=CalibrateProcess(F4;C3:C1306;1)`}
- This command is used to perform the calibration.
- In the Black-Karasinski formula, the parameter *a* corresponds to the parameter α (Mean Reversion Speed), the parameter *b to* the parameter μ (Mean Reversion Level = trend level (long-term trend)), and the parameter *sigma* to the parameter σ (Diffusion = Volatility).
- In the next step, we define the forecast horizon for the upcoming simulation (cells `F13:F25`). This covers one trading year.
- For simulation, we use the supplementary function `BK_path`, which uses the parameters in the same order as returned by the `CalibrateProcess` function.
- Again, it is important that we first completely select the output range of the calibration (cells `G13:G25`) and make sure that these cells remain selected while entering the Risk-Kit function.
- Then we enter the following Risk-Kit function in the cell window: `=BK_path (C3,F7,G7,H7,F13:F25)`
- We end the command with a curly bracket by holding down the `CTRL+ALT` keys and then pressing `ENTER`. In the cell window the following will appear: {`=BK_path(C3;F7;G7;H7;F13:F25)`}
- The stochastic values for the oil price are now calculated. These serve as the basis for the Monte Carlo simulation (Fig. 15).

- It is important to note that, it is convenient to have the historical price data in descending order of the date so that the most recent price at which the simulation begins is immediately.
- For the simulation, the parameter `DeltaT` must be selected as per the calibration in the previous step, since the process parameters are scaled accordingly. The `times` at which the simulated process is evaluated must be expressed as multiples of `DeltaT`. Thus, if `DeltaT` = 1 represents daily data available for

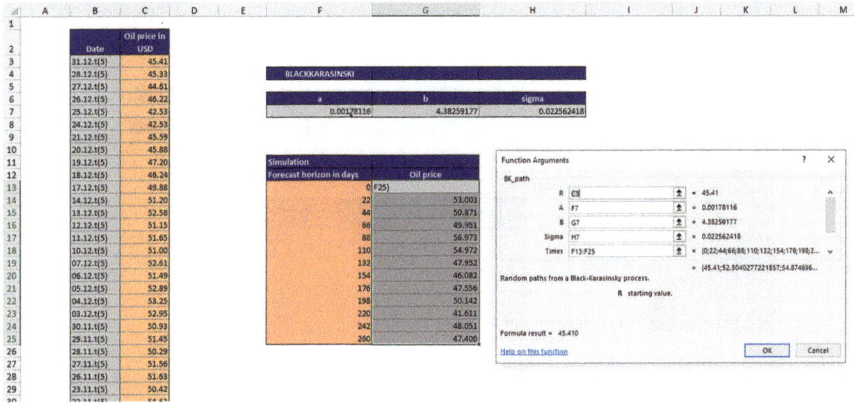

Fig. 15 Function parameters of the Black-Karasinski process

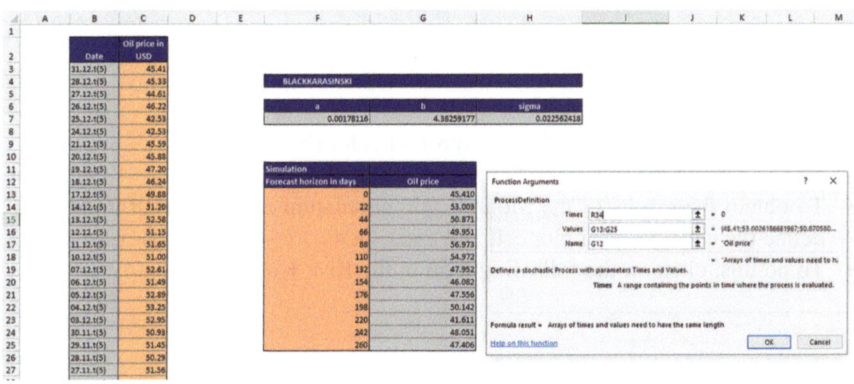

Fig. 16 Process definition of the Black-Karasinski process

5 working days per week, a forecast over a month will have a forecast horizon of 22 working days and over a year of 260 days.

- To run the Monte Carlo simulation, first enter the Risk-Kit function `ProcessDefinition` with the required values in the cell `I15` `I15=ProcessDefinition(F13:F25,G13:G25,G12)` (Fig. 16)
- Then cell `I15` is defined as the process cell of the simulation. To do this, click on the following icon in the Risk-Kit toolbar.

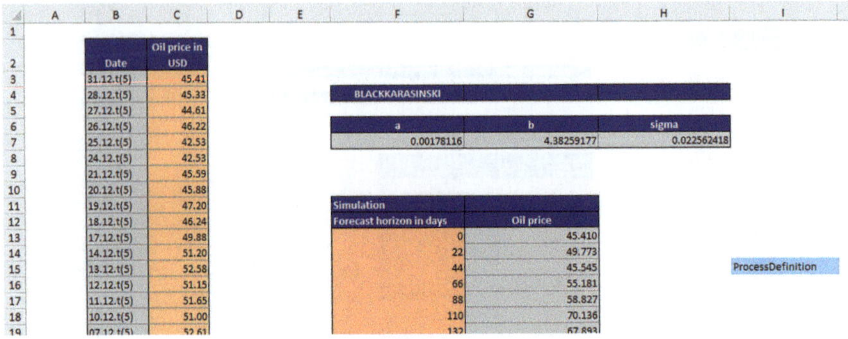

Fig. 17 Process cell of the simulation

- The cell then gets a light blue background color (Fig. 17).

- To obtain the results of the Monte Carlo simulation in the form of a graph, we define the cell `I16` as a plot cell and enter there: `I16=PlotProcess(I15)`
- To do this, click on the following icon in the Risk-Kit toolbar.

- The cell then gets an orange background color (Fig. 18).

- Before you start the Monte Carlo simulation with the icon , make sure that the  following settings have been selected (Fig. 19).

- Now you can start the Monte Carlo simulation. As a result, you receive a complete statistical evaluation of the simulation and a graphical representation of the paths. This, in turn, serves as a starting point for the risk analysis of the simulated future values.

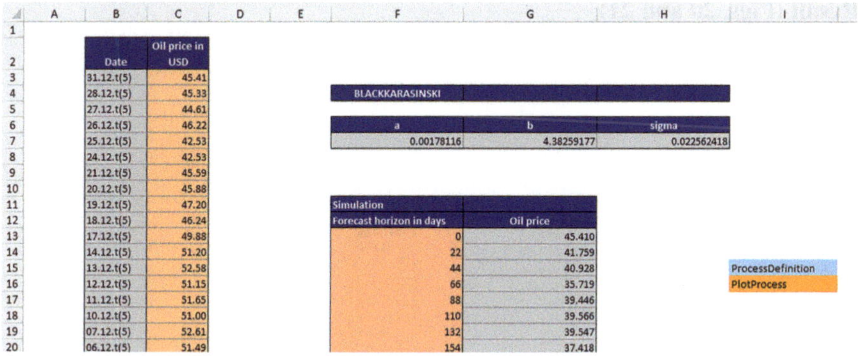

Fig. 18 Plot cell of the simulation

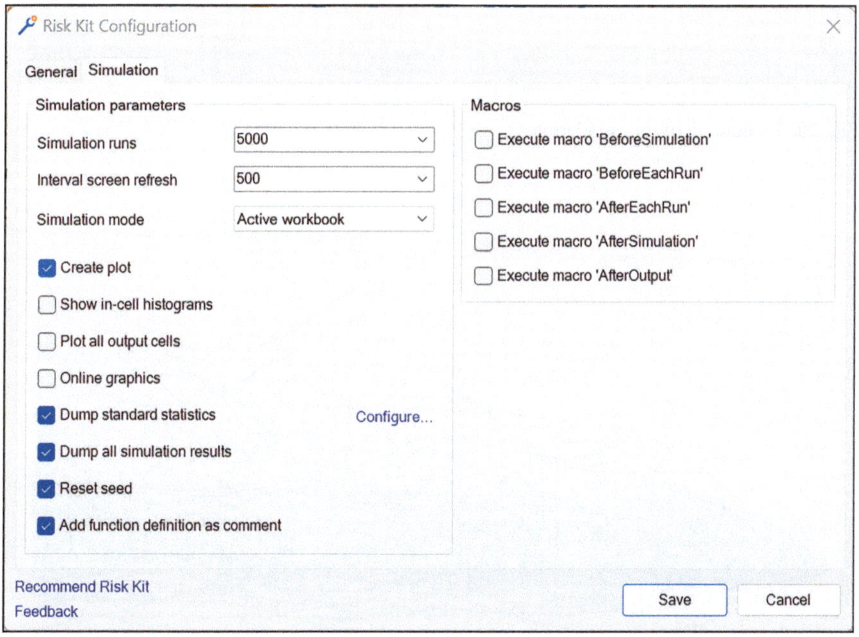

Fig. 19 Risk kit configuration

Result (Figs. 20 and 21)

	A	B	C	D	E
1	Number of Simulation Runs	10000			
2	Simulation Time	09:55.25 sec.			
3					
4					
5					
6		Out: Oil price (time point: 0)	Out: Oil price (time point: 22)	Out: Oil price (time point: 44)	Out: Oil price (time point: 66)
7	Average	45.41	46.73288508	47.91048791	49.12006042
8	Standard Deviation	9.06698E-12	4.838337442	6.922559387	8.611857639
9	Variance	8.22101E-23	23.4095092	47.92182847	74.16409199
10	Skewness	1.000150029	0.301194883	0.445197884	0.539497024
11	Kurtosis	-2.00040012	0.186114385	0.418520384	0.615300751
12	Coefficient of Variation	1.99669E-13	0.103531751	0.144489436	0.175322619
13	Mode	45.41	47.36607263	46.89814185	46.96069434
14	Mean absolute deviation	9.06653E-12	3.854861017	5.480767929	6.828933394
15	Range	0	37.78737322	64.46872789	81.28006819
16	Range5_95	0	15.71201106	22.60402228	28.21952857
17	Minimum	45.41	31.87324961	25.62346165	25.01507593
18	Expected Tail <= 0.1%	45.41	32.90067625	29.09273051	27.21843762
19	0.01% - Quantile	45.41	31.87324961	25.62346165	25.01507593
20	0.02% - Quantile	45.41	32.21935107	28.48151705	26.12584362
21	0.03% - Quantile	45.41	32.47875136	28.50250213	26.79078281
22	0.04% - Quantile	45.41	32.52457287	28.64224463	26.89394414
23	0.05% - Quantile	45.41	32.88460418	28.98567952	27.21960034
24	0.06% - Quantile	45.41	33.10788406	29.55795498	27.71652653
25	0.07% - Quantile	45.41	33.3260671	30.15154973	27.93412637
26	0.08% - Quantile	45.41	33.45545888	30.15461213	27.99813914
27	0.09% - Quantile	45.41	33.55268751	30.40752833	28.1784151
28	0.1% - Quantile	45.41	33.5841359	30.42025497	28.31192222

Fig. 20 Evaluation of the simulation

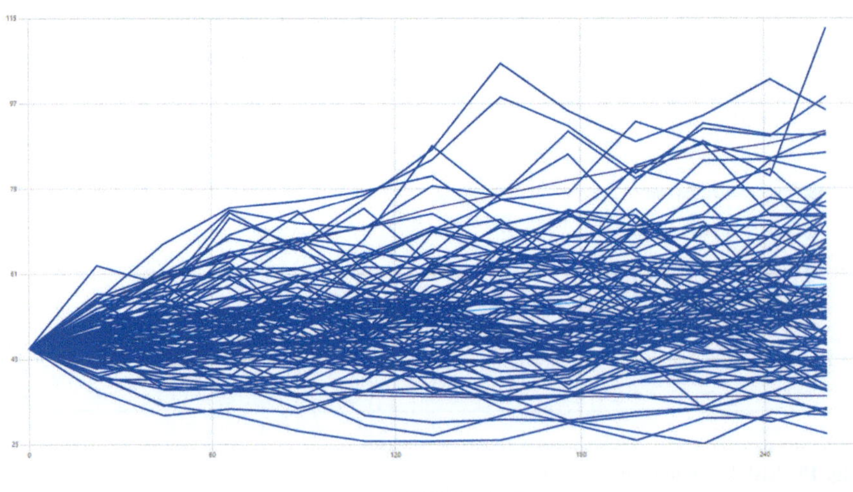

Fig. 21 Graphical representation of the simulation paths

Literature and References to the Excel Tool

Ernst, D., Häcker, J. (2021): Risk management in business step by step, UTB, Course 1—Assignment 16.

See Excel file `Case Study Risk Management Part 1`, Excel worksheet `Black-Karasinski`.

Part II
Course 2: Quantitative Instruments in Risk Management

After successfully finishing Course 1, we will proceed with Course 2.

Course 2 covers the following topics:

- Course unit 1: Different types of Value at Risk and lower partial moments and extreme value theory
- Course unit 2: Determination of portfolio risks
- Course unit 3: Hedging of hedgeable risks and modelling of non-hedgeable risks

Part II
Course 2: Quantitative Instruments in Risk Management

[illegible] Course 2 covers the following topics:

- [illegible]
- [illegible]
- [illegible]

Course Unit 1: Different Types of Value at Risk and Lower Partial Moments and Extreme Value Theory

Assignment 1: Calculation of the Value at Risk with a Discrete Probability Distribution

Assignment

- Calculate the Value at Risk for the oil price retrospectively from 31.12.2018 for the past 5 years. Please proceed with the following steps:
  - First, calculate the Value at Risk for a confidence level of 95% for 1 day based on the discrete returns of the oil price.
  - Then calculate the Value at Risk for a confidence level of 95% for 1 day based on the oil price per barrel.
  - Calculate the Value at Risk for a confidence level of 95% for 1 day based on an investment volume of 100,000 barrels of oil.
- Perform the same calculations for a 99% confidence level.
- Explain the differences in the results and give reasons for your statements.
- Create a diagram and show how the Value at Risk can be determined graphically.

Content

The *Value at Risk* (VaR) is one of the most important risk measures in financial practice. The Value at Risk (VaR) describes the minimum value that will not be exceeded with a given probability p within a certain period of time. In purely technical terms, the Value at Risk is determined as the quantile of a probability distribution or the distribution of a sample. The Value at Risk can thus assume both positive and negative values.

In practice, the Value at Risk is often used as a risk measure to determine the highest expected loss (loss potential of a certain scenario). The Value at

(continued)

D. Ernst, J. Häcker, *Corporate Risk Management*, Springer Texts in Business and Economics, https://doi.org/10.1007/978-3-031-53126-2_5

Risk thus describes the maximum loss that will not be exceeded with a certain probability (about 95% or 99%), within a certain period or holding period. The Value at Risk thus denotes a loss, i.e. a negative performance. Nevertheless, it is expressed as an absolute amount. The negative sign is omitted when stating the amount of Value at Risk, but is expressed by the word "loss." If no loss is obtained when calculating the Value at Risk with the specified probability, the Value at Risk is set to zero.

The Value at Risk (VaR) in its basic form is also called absolute Value at Risk. The *absolute Value at Risk* is an absolute risk measure, as it measures risk as an absolute loss potential and is not relative to another risk measure such as the expected value. The absolute Value at Risk thus assumes an expected value of zero. Since in reality, the expected value will rarely be zero, the absolute Value at Risk in risk management is not very effective. It is better to define the Value at Risk as a measure of risk dependent on the situation, which depends on the level of the expected value. This requirement is met by the *relative Value at Risk*. Since the use of the absolute Value at Risk, as explained, makes little sense, in practice the relative Value at Risk is often calculated directly and, for reasons of simplification, referred to as Value at Risk. However, we distinguish here between absolute and relative Value at Risk and start with the calculation of the absolute Value at Risk in its basic form.

When determining the Value at Risk using historical data, returns (relative measures) are more suitable than measures calculated in monetary units (absolute measures). This is because returns tend to have a constant expected value and a constant variance compared to absolute values. In our example of the oil price, it had a price of US$107.26 on 20.06.2014 and US$26.21 on 11.02.2016. It can be seen that price fluctuations will be higher at a price of US $100 than at a price of US$25. For the percentage changes in yields, a more uniform level can be assumed.

When determining the Value at Risk in the presence of a discrete probability distribution, the case may arise that 5% of the values considered do not result in a round number but in a decimal number. For example, as in the following example, 5% of 1303 values is 65.15 values. This means that the Value at Risk is between the 65th and 66th values. One way to determine the value of the 5% quantile is the *conservative approach*. Here, a high level of risk disclosure is preferred in order to provide additional security that the losses do not ultimately become higher than the reported Value at Risk. In this case one would round off and, as in the following example, take the 65th value as Value at Risk. Five percent of the values are less than or equal to the 65th value, i.e., −3.78%. We follow the conservative approach in this case study. An alternative is the *non-conservative approach*. Here, the value of the 5% quantile is rounded up so that the 66th value is used. This is also justified, since

(continued)

at least 5% of the values (here 5.07%) are smaller or equal to the value of –3.76%. Ninety-five percent of the values are greater than or equal to the 66th value. The non-conservative approach results in a lower Value at Risk. This approach is preferred by banks in particular, as the amount of capital required for banking supervisory purposes depends on the Value at Risk.

If the Value at Risk is calculated using a discrete probability distribution, it applies to the time unit of the return. If, as in the present example, the return is a discrete, daily return, then the calculated Value at Risk is valid for 1 day. An upscaling to weekly, monthly, quarterly, or annual values for the Value at Risk is not possible, if there is a discrete probability distribution. As we will see, this is only possible if there is a continuous probability distribution. The risk when purchasing commodities, e.g. crude oil, lies in sharp price increases that lead to higher product prices. If, on the other hand, raw materials have already been purchased, the risk lies in price falls, as in this case the products must be offered at a higher price. In this assignment of the value at risk and all other assignments, we consider the risk of price falls, i.e. the loss from an investment in crude oil.

Important Formulas

The formula for calculating the Value at Risk as return for a discrete probability distribution is:Formally speaking, the value at risk is the absolute value of the quantile Q_α. The formula for calculating the value at risk as a return with a discrete probability distribution is as follows:

$$\boxed{\mathrm{VaR}_{\mathrm{discr.return},\alpha}(r^d) = \mid Q_\alpha(r^d) \mid} \tag{1}$$

$\mathrm{VaR}_{\mathrm{discr.\ return},\ \alpha}$ = Value at risk at the α quantile
$Q_\alpha = \alpha$ quantile

The formula for calculating the Value at Risk in monetary units for a discrete probability distribution is:

$$\boxed{\mathrm{VaR}_{\mathrm{Monetary\ units},\alpha}(r^d) = \mid Q_\alpha(r^d) \cdot \mathrm{Price} \mid} \tag{2}$$

$\mathrm{VaR}_{\mathrm{Monetary\ units},\ \alpha}$ = Value at risk at the α quantile in monetary units
Price = Price per unit

Workbook: `Case Study Risk Management PHARMA GROUP` ➲ Worksheet: `VaR discr. returns (1)`

Calculation of the negative quantile of the distribution at confidence level p in Excel:

Excel example: `K11=MIN(VLOOKUP(K10;F6:H1308;3;0);0)`

Calculation of the Value at Risk as return with a discrete probability distribution in Excel:

Excel example: `K12=ABS(K11)`

Calculation of the Value at Risk in monetary units with a discrete probability distribution in Excel:

Excel example: `K16=ABS(K14*K11)`

Procedure

- First, the discrete daily returns are calculated from the oil prices (column `D`).
- The discrete daily returns are then sorted in column `H` in ascending order of size.
- To do this, select the cells `G6:H1308`, in which the date and values are noted.
- Then click on `Data` ➲ `Analysis` ➲ `Sort and Filter` ➲ `Sort` and enter as "Column" `H` "Sort by" `Values` at "Order" `Sort Smallest to Largest`.
- Then the confidence level p (cells `K7` and `N7`) and the alpha α (cells `K8` and `N8`) are determined. The confidence level p is the probability with which a loss will not be undercut. Alpha α is the probability with which the values fall below the Value at Risk.
- The next step in determining the Value at Risk is to determine the $\alpha\%$ of smallest values for a discrete probability distribution. In our example, these are the 5% of the smallest values. With 1303 available returns, this is 65.15 values. In this case, the result should be rounded down to 65 and the corresponding 65th value of the ordered list in column `I` should be selected. This is done for the 5% quantile with the function `K10=ROUNDDOWN((K8)*COUNT(H6:H1308);0)`. In our example, 5% of the smallest values are 65 set of values.
- The return on the 65th lowest value is then determined. This is done for the 5% quantile with the function `K11=MIN(VLOOKUP(K10;F6:H1308;3;0);0)`. The calculated return is −3.78%. Since the Value at Risk for discrete returns is always a positive value, it is calculated with the function `K12=ABS(K11)`. This means that at least 5% of the returns of the given distribution function (in our example 66/1302 = 5.07%) have values less than or equal to −3.78%, or: 95% of the values (in our example 1237/1302 = 95.01%) are greater than or equal to −3.78%. At 95% probability, the loss will not be greater than 3.78% the next day. If the calculated return had been positive, the Value at Risk would have been zero.
- Finally, the Value at Risk is calculated in monetary units `K16=ABS(K14*K11)`. It amounts to EUR 1.72. Furthermore, the Value at Risk for the investment volume V is also calculated `K20=K16*K18`. This amounts to EUR 171.570. With a probability of 95%, the loss from the investment will not exceed the value of EUR 171.570 within the next day.

Result (Fig. 1)

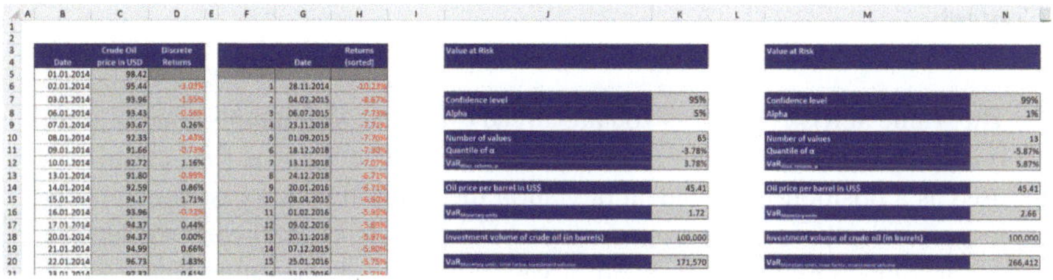

Fig. 1 Determination of the VaR with a discrete probability distribution

Literature and References to the Excel Tool

Ernst, D. (2020): Corporate Risk Management, Course 2—Assignment 1.

Standard literature:

Hopkin, P. (2018): Fundamentals of Risk Management: Understanding, Evaluating and Implementing Effective Risk, 5. Edition, New York.

Hull, J.C. (2018): Risk Management and Financial Institutions, 5. Edition, Hoboken (New Jersey).

See Excel file Case Study `Corporate Risk Management`, Excel worksheet `VaR discr. returns (1)`.

Assignment 2: Calculation of the Relative Value at Risk (Deviation Value at Risk) for a Discrete Probability Distribution

Assignment

- Calculate the relative Value at Risk for the oil price from 31.12.2018 retroactively for the last 5 years. Proceed with the following steps:
  - First, calculate the relative Value at Risk for a confidence level of 95% for 1 day based on the discrete returns of the oil price.
  - Then calculate the relative Value at Risk for a confidence level of 95% for 1 day based on the oil price per barrel.
  - Lastly, calculate the relative Value at Risk for a confidence level of 95% for 1 day based on an investment volume of 100,000 barrels of oil.
- Perform the same calculations for a 99% confidence level.
- Explain the differences in the results and give reasons for your statements.
- Create a diagram and show how the relative Value at Risk can be determined graphically.

Content

The Value at Risk (*VaR*), as you are already familiar with it in its general form, is a measure of risk that depends on the situation. The Value at Risk can also be used as a position-independent deviation measure, in which case it is referred to as *relative Value at Risk* or *deviation Value at Risk.* As already mentioned, the absolute Value at Risk is not a very suitable risk measure. Therefore, in practice, the relative Value at Risk is often referred to as Value at Risk.

The absolute Value at Risk can be transferred to the relative Value at Risk by centering the position-dependent Value at Risk on the expected value of the random variable. In this definition, the relative Value at Risk is calculated from the difference between the maximum loss at the confidence level $1-\alpha$ (VaR) and the expected value of the distribution. The relative Value at Risk is, therefore, a key figure that shows the extent of possible (negative) deviations from the plan.

Important Formulas

The formula for calculating the relative Value at Risk as return with a discrete probability distribution is:

$$\boxed{\mathrm{RVaR}_{\text{discr.return},\alpha}(r^d) = \mid Q_\alpha(r^d) - E(r^d) \mid} \tag{3}$$

$\mathrm{RVaR}_{\text{discr. return},\,\alpha}$ = Relative Value at Risk
$Q_\alpha = \alpha$ quantile
E = Expected value

The formula for calculating the relative Value at Risk in monetary units for a discrete probability distribution is:

$$\boxed{\mathrm{RVaR}_{\text{monetary units},\alpha}(r^d) = \mathrm{RVaR}_{\text{disctr.returns},\alpha}(r^d) \cdot \text{Price}} \tag{4}$$

$\mathrm{RVaR}_{\text{monetary units},\,\alpha}$ = Relative Value at Risk in monetary units
Price = Price per unit

Workbook: `Case Study Risk Management Part 1 PHARMA GROUP` ➲ Worksheet: `VaR discr. returns (1)`

Calculation of the relative Value at Risk as return with a discrete probability distribution in Excel:

Excel example: `K25=ABS(K11-AVERAGE(H6:H1308))`

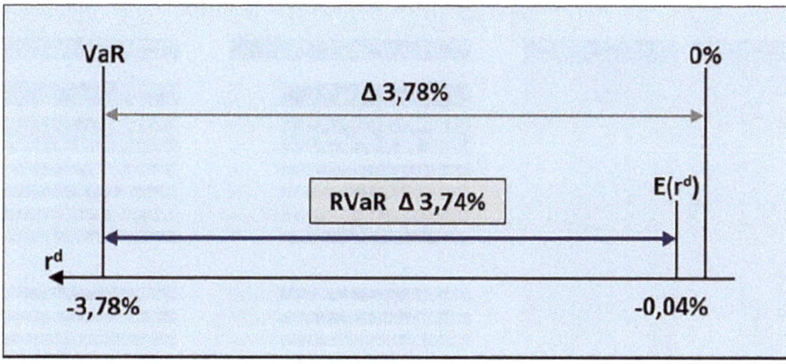

Fig. 2 Representation of the relative Value at Risk

Calculation of the relative Value at Risk in monetary units with a discrete probability distribution in Excel:

Excel example: `K27=K25*K14`

Procedure

- Based on the already calculated 5% quantile of −3.78%, the relative Value at Risk is calculated by adding the Value at Risk to the expected value (mean value) of the daily, discrete returns of the oil price and calculating the deviation from the expected value. Since the expected value has a negative value of −0.04%, the density function shifts to the left. The relative Value at Risk is 3.74%. With the shift of the density function to the left, the limit, with which a probability of *p* will not be underestimated, also shifts to the left. In this example, this explains the higher relative Value at Risk compared to the absolute Value at Risk.
- The next step is to calculate the relative Value at Risk in monetary units by multiplying the relative Value at Risk for discrete returns by the investment volume *V* and calculating the absolute value for the result. The relative Value at Risk in monetary units is EUR 1.70 and for the investment volume EUR 170.028 (Fig. 2).

Result (Fig. 3)

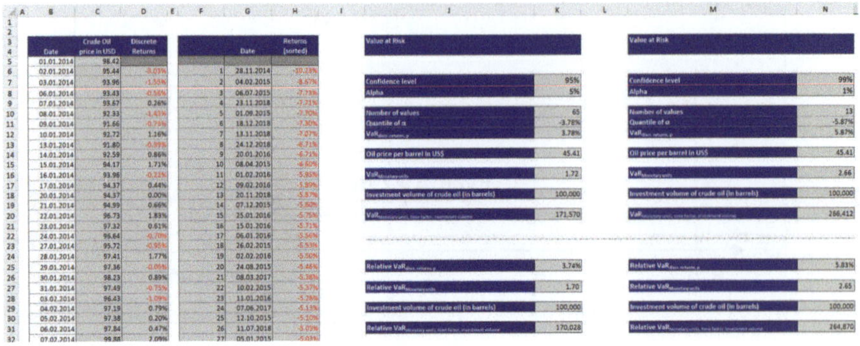

Fig. 3 Determination of the Relative VaR for a discrete probability distribution

Literature and References to the Excel Tool

Ernst, D. (2020): Corporate Risk Management, Course 2—Assignment 2.

Standard literature:

Hopkin, P. (2018): Fundamentals of Risk Management: Understanding, Evaluating and Implementing Effective Risk, 5. Edition, New York.

Hull, J.C. (2018): Risk Management and Financial Institutions, 5. Edition, Hoboken (New Jersey).

See Excel file Case Study `Corporate Risk Management`, Excel worksheet `VaR discr. return (1)`.

Assignment 3: Calculation of the Conditional Value at Risk or Expected Shortfall for a Discrete Probability Distribution

Assignment

- Calculate the discrete Conditional Value at Risk for the oil price from 31.12.2018 retroactively for the last 5 years. Proceed in the following steps:
  - First, calculate the conditional Value at Risk for a confidence level of 95% for 1 day based on the discrete returns of the oil price.
  - Then, calculate the conditional Value at Risk for a confidence level of 95% for 1 day based on the oil price per barrel.
  - Calculate the conditional Value at Risk for a confidence level of 95% for 1 day based on an investment volume of 100,000 barrels of oil.
- Perform the same calculations for a 99% confidence level.
- Explain the differences in the results and give reasons for your statements.

- Create a diagram and show how the conditional Value at Risk can be determined graphically.

Content

The *conditional Value at Risk* or the *expected shortfall* indicates the losses exceeding the Value at Risk and determines their average amount. The conditional Value at Risk or expected shortfall is also known as the conditional tail expectation or expected tail loss.

The Conditional Value at Risk or the Expected Shortfall overcomes the disadvantages of the Value at Risk, which gives no insight into what happens if the losses exceed the limit set by the Value at Risk. The conditional beyond this limit. Furthermore, the conditional Value at Risk, unlike the Value at Risk, fulfils all the conditions for a coherent measure of risk; in particular, the conditional Value at Risk is subadditive.

The Value at Risk and conditional Value at Risk are closely related. While the Value at Risk answers the question of which loss is only exceeded in a maximum of 100*α%, the conditional Value at Risk shows how high the expected loss is in 100*α% of the cases with highest losses.

Important Formulas

The formula for calculating the Conditional Value at Risk as return with a discrete probability distribution is:

$$\boxed{\text{CVaR}_{\text{discr.return},\alpha}(X) = \mid E(r^d) < Q_\alpha(r^d) \mid} \tag{5}$$

$\text{CVaR}_{\text{discr. return},\ \alpha}$ = Conditional Value at Risk

The formula for calculating the conditional Value at Risk in monetary units for a discrete probability distribution is:

$$\boxed{\text{CVaR}_{\text{monetary units},\alpha}(X) = \text{CVaR}_{\text{discr.Rendite},\alpha}(r^d) \cdot \text{Price}} \tag{6}$$

$\text{CVaR}_{\text{monetary units},\ \alpha}$ = Conditional Value at Risk in monetary units
Price = Price per unit

Workbook: `Case Study Risk Management Part 1 PHARMA GROUP` ➲ Worksheet: `VaR discr. returns (1)`

Calculation of the conditional Value at Risk (expected shortfall) as return with a discrete probability distribution in Excel:

Excel example: `K36=ABS(AVERAGE(H6:H69))`

Calculation of the conditional Value at Risk (expected shortfall) in monetary units for a discrete probability distribution in Excel:

Excel example: `K38=ABS(K36*K14)`

Procedure

- Based on the discrete returns sorted upwards in the column D, the 64 returns that fall within the 5% quantile are selected. For these 64 values, the mean value is calculated and from this the absolute value. The conditional Value at Risk shows that the expected loss in 5% of the cases with most losses is 5.17%.
- In the next step, the conditional Value at Risk (expected shortfall) is calculated in monetary units. This amounts to EUR 2.35. With an investment volume of 100,000 barrels of crude oil, the conditional Value at Risk is EUR 234.836.

Result (Fig. 4)

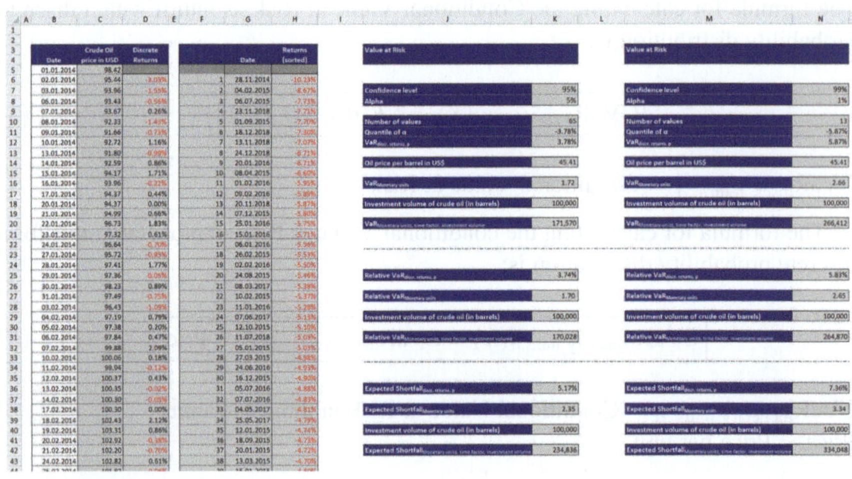

Fig. 4 Determination of the conditional VaR (expected shortfall) with a discrete probability distribution

Literature and References to the Excel Tool

Ernst, D. (2020): Corporate Risk Management, Course 2—Assignment 3.

Standard literature:

Hopkin, P. (2018): Fundamentals of Risk Management: Understanding, Evaluating and Implementing Effective Risk, 5. Edition, New York.

Hull, J.C. (2018): Risk Management and Financial Institutions, 5. Edition, Hoboken (New Jersey).

See Excel file Case Study `Corporate Risk Management`, Excel worksheet `VaR discr. returns (1)`.

Assignment 4: Calculation of the Value at Risk with a Continuous Probability Distribution

Assignment

- Calculate the Value at Risk for the oil price from 31.12.2018 retrospectively for the last 5 years. Proceed as follows:
  - First, calculate the Value at Risk for a confidence level of 95% for 1 day based on the continuous returns of the oil price.
  - Then, calculate the Value at Risk for a confidence level of 95% for 1 day based on the oil price per barrel.
  - Calculate the Value at Risk for a confidence level of 95% for 1 day based on an investment volume of 100,000 barrels of oil.
  - Calculate the Value at Risk for a confidence level of 99% for 1 day based on the continuous returns of the oil price.
  - Then, calculate the Value at Risk for a confidence level of 99% for 1 day based on the oil price per barrel.
  - Then, calculate the Value at Risk for a confidence level of 99% for 30 days based on the oil price per barrel.
  - Finally, calculate the Value at Risk for a confidence level of 99% for 30 days based on an investment volume of 100,000 barrels of oil.
- Explain the differences in the results and give reasons for your statements.
- Create a diagram and show how the Value at Risk can be determined graphically.

Content

In the case of a continuous probability distribution, the determination of the *Value at Risk* changes only slightly compared to a discrete probability distribution. Here too, the α quantile of the distribution is determined, i.e., the value below which 100*α% of the values are. In contrast to the discrete probability distribution, no historical data is used as a basis for the distribution, but a continuous distribution is assumed.

The most important continuous distribution is the normal distribution. The determination of the quantiles and thus the calculation of the Value at Risk is specifically easy in the case of a normal distribution. The normal distribution is completely determined by the expected value μ and the standard deviation σ, so that if these parameters are known, the Value at Risk for the relevant quantile can be calculated easily. We use the continuous returns as a measure of return here. In a normal distribution, the Value at Risk is automatically calculated as the relative Value at Risk. No distinction is made here between the absolute Value at Risk and the relative Value at Risk.

When calculating the Value at Risk, it should be noted that the time horizon of the Value at Risk and the time dimension of the parameters μ and σ must be identical. If the Value at Risk is calculated for 1 day, as in our example, μ and σ must also be stated as the expected value and the standard deviation of the daily return. If the time horizon of the Value at Risk and the parameter do not match, the expected value μ and the standard deviation σ must be rescaled to the time horizon of the Value at Risk. It is assumed that the parameters μ and σ are constant over time. A rescaling is only possible with continuous returns as a data basis. This is not possible with discrete returns.

Important Formulas

If the continuous return r^c of a financial product can be modelled by an $N(\mu, \sigma)$ distribution, then the following formula applies to the Value at Risk at a confidence level $p = 1 - \alpha$:

$$\boxed{\text{VaR}_{\text{contin.return},\alpha}(r^c) = |\ \sigma_T\, z_\alpha + E(r^c)\ |} \tag{7}$$

$\text{VaR}_{\text{contin. return},\ \alpha}$ = Value at Risk for continuous returns
σ_T = Standard deviation
z_α = α-quantile of the standard normal distribution
$E = \mu$ = Expected value

The formula for calculating the Value at Risk in monetary units for a normal distribution is:

$$\boxed{\text{VaR}_{\text{monetary units},\alpha}(r^c) = \text{VaR}_{\text{contin.return},\alpha}(r^c) \cdot \text{Price}} \tag{8}$$

$\text{VaR}_{\text{monetary units},\ \alpha}$ = Value at risk in monetary units
Price = Price per unit

If the Value at Risk is to be calculated for a period other than 1 day, e.g., for 30 days, the expected value μ, and the standard deviation σ must be rescaled to the time horizon of the Value at Risk. We assume that the parameters refer to a period of duration *T*, in our example daily. Now the Value at Risk is to be calculated for a period with the duration *N*T*, e.g., 30 days. The conversion of the expected value μ is simple and is done with the formula:

$$\boxed{\mu_{N \cdot T} = N \cdot \mu_T} \tag{9}$$

N = Time viewed
T = Time factor

If the returns between the sub-periods are independent, the standard deviation is σ:

$$\boxed{\sigma_{N \cdot T} = \sqrt{N} \cdot \sigma_T} \tag{10}$$

Inserting the rescaled parameters in the Value at Risk formula yields the following equation for the rescaled Value at Risk:

$$\boxed{\text{VaR}_{\text{contin.return},\alpha}(r^c) = \left| \sqrt{N} \cdot \sigma_T \cdot z_\alpha - N \cdot \mu \right|} \tag{11}$$

Workbook: `Case Study Risk Management Part 1-PHARMA GROUP` ➲
Worksheet: `VaR continuous returns (1)`

Calculation of the negative quantile of the distribution at confidence level *p*, assuming a normal distribution in Excel:

Excel example: `H10=NORM.INV(H9;0;1)`

Calculation of the continuous daily volatility in Excel:

Excel example: `H12=E1308/SQRT('Assumptions General'!C217)`

The 5% quantile is calculated in Excel:

Excel example: `H14=H10*H12`

The expected value of the continuous return is calculated in Excel using the `AVERAGE` function:

Excel example: `H16=AVERAGE(D6:D1308)`

The Value at Risk is calculated in Excel with:

Excel example: `H18=ABS(H10*H12-H16)`

Taking into account the time dimension, the Excel formula for the Value at Risk is:

Excel example: `H27=ABS(H25*H10*H12-H24*H16)`

Taking into account the time dimension, the price, and the investment volume, the Excel formula for the Value at Risk:

Excel example: `H32=H28*H30`

Procedure

- First, the continuous return in column `D` and the annualized volatilities are calculated based on 30 days in column `E` and 250 days in column `F`.
- Then, in the cell `H10=NORM.INV(H9,0,1)` the value of the standard normal distribution for the 5% quantile is calculated. The result is −1.645.

- The daily volatility is then calculated based on the annualized 250-day volatility in the cell H12= `E1308/SQRT('Assumptions General'!C217)`. The result is 1.97%.
- The 5% quantile is −3.24% and is calculated using the Excel formula H14=H10*IH12.
- Furthermore, the mean value of the continuous daily return is required to calculate the Value at Risk in the cell `H16=AVERAGE(D6:D1308)`. This has a value of −0.06%.
- The Value at Risk for 1 day is 3.30% and is calculated with the Excel formula in the cell `H18=ABS(H10*H12-H16)`.
- At a barrel price of US$45.41, the Value at Risk in monetary units is US$1.45 per day. The Excel formula in the cell `H22=H18*H20` is used.
- Adding the time factor and the total investment volume of 100,000 barrels, the Value at Risk is 3.18% according to the Excel formula `H27=ABS(H25*H10*H12-H24*H16)`.
- The Value at Risk in monetary units amounts to US$144,608. The Excel formula `H32=H28*H30` is used.
- The Value at Risk states that in 95% of cases, the loss from an investment in 100,000 barrels of crude oil does not exceed US$144,608 per day.

Result (Fig. 5)

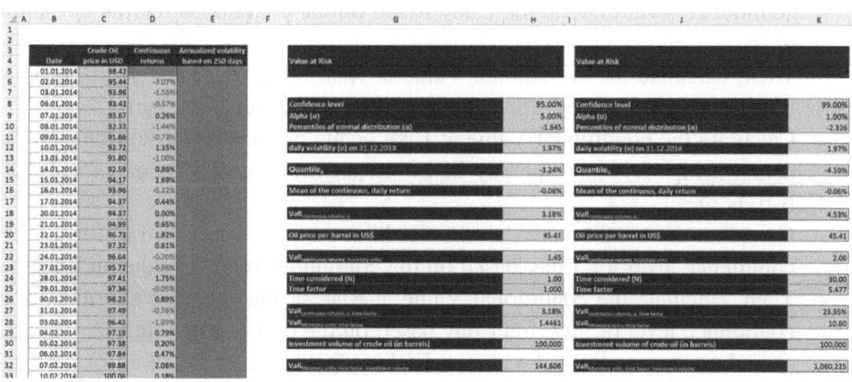

Fig. 5 Determination of the Value at Risk with a continuous probability distribution

Literature and References to the Excel Tool

Ernst, D. (2020): Corporate Risk Management, Course 2—Assignment 4.

Standard literature:

Hopkin, P. (2018): Fundamentals of Risk Management: Understanding, Evaluating and Implementing Effective Risk, 5. Edition, New York.

Hull, J.C. (2018): Risk Management and Financial Institutions, 5. Edition, Hoboken (New Jersey).

See Excel file Case Study `Corporate Risk Management`, Excel worksheet `VaR continuous returns (1)`.

Assignment 5: Calculation of the Conditional Value at Risk or Expected Shortfall for a Continuous Probability Distribution

Assignment

- Calculate the Conditional Value at Risk or the expected shortfall for the oil price from 31.12.2018 retroactively for the last 5 years. Proceed in the following steps:
  - First, calculate the conditional Value at Risk or the expected shortfall for a confidence level of 95% for 1 day based on the continuous returns of the oil price.
  - Then, calculate the conditional Value at Risk or the expected shortfall for a confidence level of 95% for 1 day based on the oil price per barrel.
  - Calculate the conditional Value at Risk or the expected shortfall for a confidence level of 95% for 1 day based on an investment volume of 100,000 barrels of oil.
  - Calculate the conditional Value at Risk or the expected shortfall for a 99% confidence level for 1 day based on the continuous returns of the oil price.
  - Then, calculate the conditional Value at Risk or the expected shortfall for a confidence level of 99% for 1 day based on the oil price per barrel.
  - Calculate the conditional Value at Risk or the expected shortfall for a confidence level of 99% for 30 days based on the oil price per barrel.
  - Lastly, calculate the conditional Value at Risk or the expected shortfall for a confidence level of 99% for 30 days based on an investment volume of 100,000 barrels of oil.
- Explain the differences in the results and give reasons for your statements.
- Create a diagram and show how the Value at Risk can be determined graphically.

Content

With a continuous probability distribution, the *conditional Value at Risk* or the *expected shortfall* is formed as the expected value of all losses that are greater than or equal to the Value at Risk.

Important Formulas

The expected value corresponds to the calculation of the integral:

$$\mathrm{CVaR}_\alpha(r^c) = \int_0^\alpha \mathrm{VaR}_u \; du \tag{12}$$

$\mathrm{CVaR}_{\text{contin. return}, \alpha}$ = Conditional Value at Risk at confidence level p for continuous returns

For a normally distributed random variable $X \sim N(\mu, \sigma)$ the following results:

$$\mathrm{CVaR}_\alpha(r^c) = \sigma \cdot \frac{\varphi(z_\alpha)}{\alpha} + \mu \tag{13}$$

$\varphi(z)$ is the density of the standard normal distribution. It is evaluated on the $(1\text{-}\alpha)$-quantile of the standard normal distribution.

By inserting the rescaled parameters into the Value at Risk formula, the following equation of the rescaled Value at Risk is obtained:

$$\mathrm{CVaR}_{(\alpha)} = \sqrt{N} \cdot \sigma_T . \frac{\varphi(Z_{1-\alpha})}{\alpha} + N \cdot \mu_T \tag{14}$$

Workbook: `Case Study Risk Management Part 1 PHARMA GROUP` ➲ Worksheet: `VaR continuous returns (1)`

The density is calculated in the Excel as follows:

Excel example: `H37=NORM.DIST(NORM.INV(H9;0;1);0;1;FALSE)` =

`=NORM.DIST(H10;0;1;FALSE)`

Calculation of the conditional Value at Risk (expected shortfall) as a return with a continuous probability distribution in the Excel:

Excel example: H38=SQRT(H24)*H12*H37/H9+H24*H16

Calculation of the conditional Value at Risk (expected shortfall) in monetary units with a continuous probability distribution in Excel:

Excel example: H39=H20*H38

Taking into account the time dimension, the Excel formula for the Value at Risk is:

Excel example: H40=(H12*SQRT(H24)*H37/H9+H24*H16)*H20*H30

Procedure

- In the first step, the density is calculated for the 5% quantile H37=NORM.DIST(H10;0;1;FALSE). It amounts to 0.1031.
- The conditional Value at Risk is then calculated for 1 day in the cell H38=SQRT(H24)*H12*H37/H9+H24*H16. The conditional Value at Risk for 1 day is 4.01%. The conditional Value at Risk can also be calculated for any other period, e.g., for 30 days. Then, the cell H24 must be adjusted accordingly.
- The conditional Value at Risk is converted into monetary units by multiplying the conditional Value at Risk for 1 day by the price of oil in barrels. The conditional Value at Risk in monetary units is US$1.82.
- For the investment volume of 100,000 barrels, which is examined here, the conditional Value at Risk is US$182,029.
- The conditional Value at Risk states that in 5% of the cases, the loss from an investment in 100,000 barrels of oil exceeds US$144,608 per day. The average loss then amounts to US$182,029.

Result (Fig. 6)

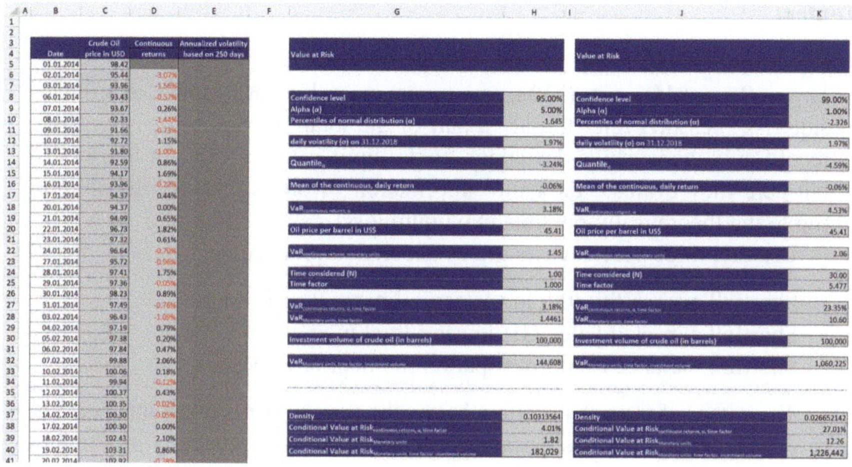

Fig. 6 Determination of the conditional VaR (expected shortfall) with a continuous probability distribution

Literature and References to the Excel Tool

Ernst, D. (2020): Corporate Risk Management, Course 2—Assignment 5.

Standard literature:

Hopkin, P. (2018): Fundamentals of Risk Management: Understanding, Evaluating and Implementing Effective Risk, 5. Edition, New York.

Hull, J.C. (2018): Risk Management and Financial Institutions, 5. Edition, Hoboken (New Jersey).

See Excel file Case Study `Corporate Risk Management`, Excel worksheet `VaR continuous returns (1)`.

Assignment 6: Calculation of Lower Partial Moments: Shortfall Probability

Assignment

Calculate the shortfall probability (0th lower partial moment or LPM_0), for the oil price from 31.12.2018 retroactively for the last 5 years.

Content

Lower Partial Moments (LPM_m measures) are risk measures that, as a downside risk measure, refer to only a part of the total probability density. They record only the negative deviations from a limit τ (target value or threshold), but here they evaluate the entire information of the probability distribution (up to the theoretically possible maximum loss).

Usually, three special cases are considered in practice:

- Shortfall probability (0th lower partial moment or LPM_0)
- Shortfall expected value (1st lower partial moment or LPM_1)
- Shortfall variance (2nd lower partial moment or LPM_2)

The extent of the risk of falling below the threshold τ is taken into account in various ways. For the shortfall probability, only the probability p of falling below the threshold plays a role. In contrast, the expected shortfall value takes into account the average level of shortfall.

The *shortfall probability* (0th lower partial moment or LPM_0)—also known as the downside probability—measures the probability that the minimum requirement level will fall below a predefined target value (threshold) τ. However, the absolute level of underachievement of the target value is not recorded. The shortfall probability or probability of default is calculated, for example, as the number of returns that fall below the target value in relation to the total number of returns. The shortfall probability thus represents the relative frequency of the target value being undershot.

To determine the shortfall probability, the minimum requirement level is defined and the probability of falling below it is determined. By contrast, with VaR as the α quantile, the shortfall probability is determined by the confidence level selected and the corresponding target value is determined. The shortfall probability and the VaR are thus inversely related.

Important Formulas

The general formula of the lower partial moments is:

$$\boxed{\mathrm{LPM}_n(\tau, r^c) = \frac{1}{T} \sum_{\substack{t=1 \\ r^c<\tau}}^{T} (\tau - r^c)^n} \tag{15}$$

LPM = Lower partial moments

τ = Required target value, also “target”, e.g. the minimum return

r^c = Continuous return

t = Index for the time

T = Time span

For the shortfall probability, only the probability P of the shortfall plays a role. The shortfall probability (LPM_0) is calculated based on the above formula as follows:

$$\boxed{\mathrm{LPM}_0(\tau, r^c) = P(r^c < \tau)} \tag{16}$$

Workbook: `Case Study Risk Management PHARMA GROUP` ➲ Worksheet: `Lower partial Moments`

The calculation of the shortfall probability in Excel is done according to:

> Excel example: `L9=COUNT(E6:E1308)/COUNT(D6:D1308)`

Procedure

- Starting from the discrete returns in column `D`, the returns X are determined, which are below the target value τ (the target value is shown in the cell `L7` and is −3.5%). This is done using the Excel formula in the cell `E6=IF(D6<$L$7; D6;"")`.
- In the next step, the shortfall probability (LPM_0) is calculated by dividing the number of values under the target by the number of observable returns as per the cell `L9=COUNT(E6:E1308)/COUNT(D6:D1308)`.

Result (Fig. 7)

Literature and References to the Excel Tool

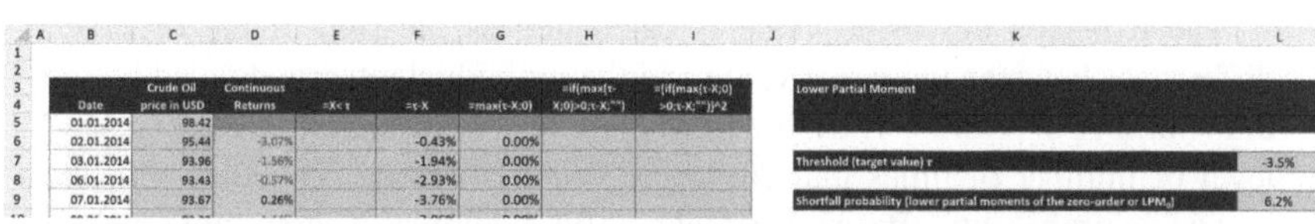

Fig. 7 Calculation of the shortfall probability (zero-order lower partial moments or LPM_0)

Ernst, D. (2020): Corporate Risk Management, Course 2—Assignment 6.
Standard literature:
Hopkin, P. (2018): Fundamentals of Risk Management: Understanding, Evaluating and Implementing Effective Risk, 5. Edition, New York.
Hull, J.C. (2018): Risk Management and Financial Institutions, 5. Edition, Hoboken (New Jersey).
See Excel file Case Study `Corporate Risk Management`, Excel worksheet `Lower Partial Moments`.

Assignment 7: Calculation of Lower Partial Moments: Shortfall Expected Value

Assignment

Calculate the shortfall expected value (first lower partial moment or LPM_1) for the oil price from 31.12.2018 retroactively for the last 5 years.

Content

The expected shortfall value (first lower partial moment or LPM_1), takes into account the extent to which the value falls below the specified target. It is also called "expected default," "medium default risk," "expected shortfall magnitude," or "target shortfall." At the same time, the expected shortfall can be seen as a special case of LPM_1, in which a flexible threshold value equal to the α quantile is used. The semi-standard deviation is also a special case of LPM_1.

The expected downside value is calculated as the sum of the observed differences between the target value and the individual returns, divided by the total number of returns. Thus, the LPM_1 provides information on the expected level of number of times that the target value is missed. Together with the downside probability, this gives a meaningful picture, as it shows not only the probability, but also the extent to which the target will be missed.

Important Formulas

For the shortfall expected value, the mean undercut is calculated. The calculation of the expected shortfall value in the Excel sheet is performed according to:

$$\mathrm{LPM}_1(\tau, r^c) = \frac{1}{T} \sum_{\substack{t=1 \\ r^c < \tau}}^{T} (\tau - r^c) \tag{17}$$

Workbook: `Case Study Risk Management PHARMA GROUP` ➲ Worksheet: `Lower partial Moments`

The expected shortfall value (LPM) is calculated based on the above formula as follows:

Excel example: `L11=AVERAGE(H6:H1308)`

Procedure

- Based on the target value τ of -3.5%, the difference between the return and the target value is measured. This is done using the Excel formula `F6=$L$7-D6`.
- Then, the values that are smaller than zero are determined in the cell `G6=MAX(F6;0)`. Only these values are then shown in the cell `H6=IF(G6>0;G6;"")`.
- In the last step, the expected shortfall value (LPM_1) is determined as the mean value of the deviations from the barrier `L11=AVERAGE(H6:H1308)`.
- The mean value is 1.5% and this means that in 6.2% of the cases, the losses are on average 1.5% higher than the specified barrier of -3.5% and that in 6.2% of the cases the losses are 5.0%.

Result (Fig. 8)

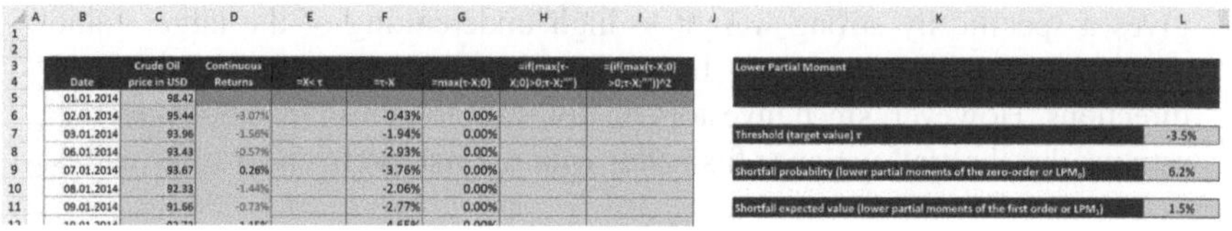

Fig. 8 Calculation of the shortfall expected value (LPM first order or LPM_1)

Literature and References to the Excel Tool

Ernst, D. (2020): Corporate Risk Management, Course 2—Assignment 7.

Standard literature:

Hopkin, P. (2018): Fundamentals of Risk Management: Understanding, Evaluating and Implementing Effective Risk, 5. Edition, New York.

Hull, J.C. (2018): Risk Management and Financial Institutions, 5. Edition, Hoboken (New Jersey).

See Excel file Case Study `Corporate Risk Management`, Excel worksheet `Lower Partial Moments`.

Assignment 8: Calculation of Lower Partial Moments: Shortfall Variance

Assignment

Calculate the second lower partial moment or LPM_2 for the oil price of 31.12.2018 retroactively for the last 5 years.

Content

The *shortfall variance* as a second lower partial moment (LPM_2) captures the dispersion of the undershoots of the target value. With the help of the shortfall variance, the larger losses are weighted more strongly than the smaller ones by squaring the under-runs of the target value. The downside variance thus measures the average squared negative deviation from the target value.

The shortfall variance has similarities to the semi-variance, which is why it is sometimes called target semi-variance. It also has similarities to the variance, although here the deviations are calculated by taking the target value into account and not the mean value of the distribution. The Shortfall variance gives a specifically strong weight to high undercutting of the target value, whereas the variance penalises strong deviations from the mean value in both directions. However, since investors do not want to punish the above-average returns, the shortfall variance fits better with the risk preferences of investors.

Important Formulas

For the shortfall expected value, the mean undercut is calculated. The calculation of the expected shortfall value in the Excel is performed according to:

$$\boxed{\mathrm{LPM}_2(\tau, r^c) = \frac{1}{T}\sum_{\substack{t=1 \\ r^c<\tau}}^{T} (\tau - r^c)^2} \quad (18)$$

Workbook: `Case Study Risk Management PHARMA GROUP` ➲ Worksheet: `Lower partial Moments`

The shortfall variance (LPM_2) is calculated based on the above formula as follows:

> Excel example: `L13=AVERAGE(I6:I1308)`

Procedure

- Starting from the target value τ with a value of -3.5% in the cell L7, the difference between return and target value is measured and squared. This is done using the Excel formula `F6=$L$7-D6`.
- Then, the values less than zero are determined in the cell are G6=MAX(F6;0)and squared `I6=IF(H6="";"";H6^2)`.
- In the last step, the shortfall variance (LPM_2) is determined as the mean value of the squared deviations from the barrier in the cell `L13=AVERAGE(I6:I1308)`.

Result (Fig. 9)

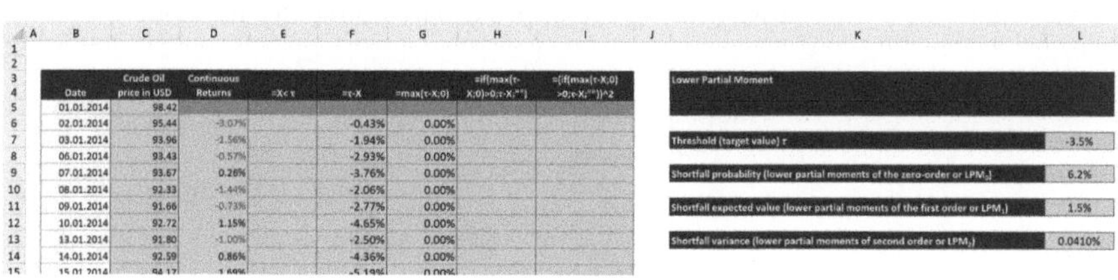

Fig. 9 Calculation of the shortfall variance as a second-order LPM (LPM_2)

Literature and References to the Excel Tool

Ernst, D. (2020): Corporate Risk Management, Course 2—Assignment 8.

Standard literature:

Hopkin, P. (2018): Fundamentals of Risk Management: Understanding, Evaluating and Implementing Effective Risk, 5. Edition, New York.

Hull, J.C. (2018): Risk Management and Financial Institutions, 5. Edition, Hoboken (New Jersey).

See Excel file Case Study `Corporate Risk Management`, Excel worksheet `Lower Partial Moments`.

Assignment 9: Value at Risk for Non-linear Price Functions: Bonds

Assignment

You hold a bond with a remaining term of 6 years. You receive interest of EUR 45 for 5 more years, and at the end of the sixth year, you receive the nominal value of the bond of EUR 1000 and EUR 45 interest for the last year of the investment.

- Calculate the present value of the payment at a market interest rate of 4.00% and after increasing the market interest rate by 1.00–5.00%.
- Calculate the duration.
- Calculate the modified duration and perform a control calculation.
- Calculate the Duration and the Modified Duration with Excel function `DURATION` and `MDURATION`.
- Calculate the Value at Risk using the exact calculation method.
- Calculate the Value at Risk using the method of approximate calculation using Macauley duration and modified duration.
- Calculate the Value at Risk using the approximate convexity method.

Content

Non-linear price functions apply to financial products whose value/price depends non-linearly on the underlying risk factors. In the following section, we consider non-linear functions of:

- Bonds
- Options

Bonds depend non-linearly on the interest rate and stock options non-linearly on the share price.

Bonds are debt instruments that are used to obtain a loan in the capital market. We assume that you have mastered the basics of bonds (see references).

The following basic terms are relevant:

Bond	Usually a longer-term bond issued by the issuer pays interest to the buyer of the bond for the agreed term, at the end of the term the buyer gets back the amount borrowed
Coupon	Entitlement to receive the interest due
Nominal value	Value of a single bond
Nominal interest rate	Interest rate at which interest is paid on the bond
Fixed-interest bond	Bond with interest rate fixed in advance. Interest is usually paid annually or semi-annually

Here, we will look at how the Value at Risk for bonds can be calculated. The calculation of the Value at Risk follows the approaches we have already seen for discrete returns. The Value at Risk for bonds can be calculated either by an "exact calculation" (as we have already done) or by an "approximate calculation." The following three risk parameters, which are central to bonds, are used for the approximate calculation:

1. Duration (Macaulay Duration) => reinvestment risk measure
2. Modified Duration => risk measure for the interest rate risk
3. Convexity => risk measure for the interest rate risk

The *duration* or *Macaulay duration*, which Frederick R. Macaulay introduced, is a key figure that can be interpreted as a time variable. The duration reflects the average period of capital commitment of a financial investment in a fixed-interest security. It indicates the time after which the investor recovers his invested capital. The duration thus describes the reinvestment risk.

Formally, the duration can be understood as the weighted average of the dates of payments of a fixed-interest security (average capital commitment

(continued)

period), whereby the present values of interest and redemption payments in relation to the total present value are used as weightage factors. Thus, the Macauley duration is nothing more than a weighted arithmetic mean of the cash flows of a fixed-interest security.

Modified duration is a measure of interest rate risk and can be derived from duration (hence modified duration). While the duration is measured in years, the modified duration answers the question frequently asked in practice of how large the relative change in the bond price is in relation to a change in the market interest rate level. The modified duration indicates the percentage change in the bond price if the market interest rate level changes by 1 percentage point. The modified duration therefore indicates how much the total return on a bond (consisting of the repayments, coupon payments and the compound interest effect when the repayments are reinvested) changes when the market interest rate changes. Thus, it measures the price effect triggered by a marginal interest rate change and can be understood as the elasticity of the bond price in relation to the market interest rate.

The present value of bonds shows a convex curve when the interest rates change. As the modified duration only takes the first derivative—the slope—into account, it only provides useful values for small interest rate changes. The modified duration is a very cautious measure of risk, as risks are overestimated and the opportunities underestimated. Convexity is, therefore, used as a more precise measure of risk.

Convexity is a measure of risk that describes the behavior of a bond when the interest rates change. It is an extension or improvement of the modified duration. A positive convexity describes the bonds that have low price sensitivity when interest rates rise and high price sensitivity when the interest rates fall. Low price falls/losses are expected when interest rates rise, while high price increases are expected when interest rates fall. Furthermore, the greater the convexity, the more pronounced the bond's behavior.

Important Formulas

The *Macaulay duration* formula is defined as follows. For reasons of simplification, a flat yield curve with a constant annual market interest rate is assumed.

$$\boxed{D = \frac{\sum_{t=1}^{T} t \cdot \frac{CF_t}{(1+r)^t}}{P} = \frac{\sum_{t=1}^{T} t \cdot \frac{CF_t}{(1+r)^t}}{\sum_{t=1}^{T} \frac{CF_t}{(1+r)^t}}} \tag{19}$$

D = Duration
CF_t = Cash flow at the end of the period t,
r = Market interest rate

P = Price of the bond
t = Time period, here 1 year

Duration is a time variable and indicates the average period of capital commitment. The return periods are weighted by the share of the net present value of the payment in the respective period in the total net present value of all payments (= bond price). The sum of the weights is one.

An immediate change in the market interest rate Δi leads to the following recalculation of the present value. The starting point for the modified duration is again the present value of the cash flow.

$$\boxed{P(r+\Delta r)=\sum_{t=1}^{T}\frac{\mathrm{CF}_t}{(1+r+\Delta r)^t}} \tag{20}$$

Mathematically, $P(r + \Delta r)$ can also be described using Taylor expansion. The modified duration is obtained by transformation, which is defined as:

$$\boxed{\mathrm{MD}=\frac{dP}{dr}\cdot\frac{1}{P}=\frac{\sum_{t=1}^{T} t\cdot\frac{\mathrm{CF}_t}{(1+r)^{t+1}}}{\sum_{t=1}^{T}\frac{\mathrm{CF}_t}{(1+r)^t}}} \tag{21}$$

The modified duration is nothing more than the first derivative of the present value function upon interest divided by the price (present value) of the bond.

The connection between the modified duration and the Macaulay duration becomes apparent when the modified duration is multiplied by $1 + r$. Then, the discounting factors and the numerator and denominator receive the same exponents and the modified duration results in the Macaulay duration.

$$\boxed{D=\mathrm{MD}\cdot(1+r)=\frac{\sum_{t=1}^{T} t\cdot\frac{\mathrm{CF}_t}{(1+r)^t}}{\sum_{t=1}^{T}\frac{\mathrm{CF}_t}{(1+r)^t}}} \tag{22}$$

As mentioned above, convexity optimizes the results of modified duration. This is achieved by adding the second derivative in the Taylor expansion. This results in the following relationship:

$$\boxed{P(r+\Delta r)\approx P(r)+\frac{dP}{dr}\cdot\Delta r+\frac{1}{2!}\cdot\frac{d^2P}{dr^2}(\Delta r)^2} \tag{23}$$

and as a consequence

$$\boxed{\frac{P(r+\Delta r)-P(r)}{P(r)}=\frac{\Delta P}{P(r)}\approx -\mathrm{MD}\cdot\Delta r+\frac{1}{2}\cdot C\cdot(\Delta r)^2} \tag{24}$$

where C describes the convexity.

$$\boxed{C=\frac{\Delta^2 P}{\Delta i^2}\cdot\frac{1}{P}=\frac{\sum_{t=1}^{T} t\cdot(t+1)\frac{\mathrm{CF}_t}{(1+r)^{t+2}}}{P}=\frac{\sum_{t=1}^{T} t\cdot(t+1)\frac{\mathrm{CF}_t}{(1+r)^{t+2}}}{\sum_{t=1}^{T}\frac{\mathrm{CF}_t}{(1+r)^t}}} \tag{25}$$

The convexity is nothing more than the second derivative of the present value function upon interest divided by the price (present value) of the bond.

The formula for the *Value at Risk after the exact calculation* is:

$$\boxed{\mathrm{VaR}=|\,P(r+\Delta r)-P(r)\,|} \tag{26}$$

The formula for the *Value at Risk after the approximate calculation using the Macaulay duration and the modified duration* is:

$$\boxed{\mathrm{VaR}=\left|\frac{1}{1+r}\cdot D\cdot P(r)\cdot\Delta r\right|=|\mathrm{MD}\cdot P(r)\cdot\Delta r|} \tag{27}$$

The formula for the *Value at Risk after the approximate calculation using convexity* is:

$$\boxed{\mathrm{VaR}=\left|-\mathrm{MD}\cdot\Delta r\cdot P(r)+\frac{1}{2}\cdot C\cdot(\Delta r)^2\cdot P(r)\right|} \tag{28}$$

Workbook: `Case Study Risk Management PHARMA GROUP` ➲ Worksheet: `VaR non-linear functions`

The calculation of the total present value of all the payments at the current market interest rate in the numerator is done in the Excel according to:

Excel example: `C21=(C8/(1+C56)^C7+D8/(1+C56)^D7+E8/(1+C56)^E7+F8/(1+C56)^F7+G8/(1+C56)^G7+(H8+I8)/(1+C56)^H7)`

The calculation of the total present value of all the payments after increasing the market interest rate is done in the Excel according to:

Excel example: `C22=(C8/(1+C57)^C7+D8/(1+C57)^D7+E8/(1+C57)^E7+F8/(1+C57)^F7+G8/(1+C57)^G7+(H8+I8)/(1+C57)^H7)`

The calculation of the present value relevant to the duration in the meter is done in the Excel according to:

Excel example: `C27=(C7*C8/(1+C56)^C7+D7*D8/(1+C56)^D7+E7*E8/(1+C56)^E7+F7*F8/(1+C56)^F7+G7*G8/(1+C56)^G7+H7*(H8)/(1+C56)^H7)`

The duration is calculated in the Excel according to:

Excel example: `C30=C27/C28`

The duration is calculated with the Excel function `DURATION` according to:

Excel example: `C61=DURATION(C53;C54;C55;C56;C58;C59)`

The calculation of the present value relevant to the modified duration in the meter is done in the Excel according to:

Excel example: `C35=(C7*C8/(1+C56)^D7+D7*D8/(1+C56)^E7+E7*E8/(1+C56)^F7+F7*F8/(1+C56)^G7+G7*G8/(1+C56)^H7+H7*(H8)/(1+C56)^I7)`

The modified duration is calculated in the Excel according to:

Excel example: `C38=C35/C36`

The duration is calculated with the Excel function `MDURATION` according to:

Excel example: `C62=MDURATION(C53;C54;C55;C56;C58;C59)`

The calculation of the present value relevant to the convexity in the numerator is done in the Excel according to:

Excel example: `C44=(C7*D7*C8/(1+C56)^E7+D7*E7*D8/(1+C56)^F7+E7*F7*E8/(1+C56)^G7+F7*G7*F8/(1+C56)^H7+G7*H7*G8/(1+C56)^I7+H7*I7*(H8)/(1+C56)^J7)`

Convexity is calculated in the Excel according to:

Excel example: `C47=C44/C45`

The exact calculation of the VaR is done in the Excel according to:

Excel example: `C67=ABS(C22-C21)`

The approximate calculation of VaR using the Modified Duration is performed in the Excel according to:

Excel example: `C73=ABS(C21*C38*C15)`

The approximate calculation of VaR using convexity is performed in the Excel according to:

Excel example: `C74=ABS(-C38*C15*C21+0,5*C47*C15^2*C21)`

Procedure

- In the payments (cells `C7:J8`) resulting from the bond, you can see the coupon/nominal interest rate. In our example, EUR 1,000 was invested. The interest payments are EUR 45 per year, i.e., the coupon/nominal interest rate is 4.5% (cell `C13`).
- Based on the payments, the present values of the payments are calculated using the current market interest rate (cell `C14`) and the market interest rate after interest rates increase (cell `C16`) (cells `C21` and `C22`). The present value of the payments discounted at the current interest rate (cell `C21`) corresponds to the price of the bond.
- In the next step, the duration, the modified duration, and the convexity of the bond are calculated without and with the Excel functions.
- To calculate the duration, the following present value is calculated in the numerator:

C27=(C7*C8/(1+C56)^C7+D7*D8/(1+C56)^D7+E7*E8/(1+C56)^E7+F7*F8/(1+C56)^F7+G7*G8/(1+C56)^G7+H7*(H8)/(1+C56)^H7)

This present value is divided by the price of the bond (cell C28), which results in the duration (cell C30). The same result is obtained using the Excel function DURATION (cell C61), which requires the inputs from the cells C53:C59 (except C57).

Both calculations result in a duration of 5.3391, which means that the investor gets back his invested capital of EUR 1000 after 5.3391 years.

- To calculate the modified duration, the following present value is calculated in the numerator:

  C35=(C7*C8/(1+C56)^D7+D7*D8/(1+C56)^E7+E7*E8/(1+C56)^F7+F7*F8/(1+C56)^G7+G7*G8/(1+C56)^H7+H7*(H8)/(1+C56)^I7)

  This present value is divided by the price of the bond (cell C36), which results in the modified duration (cell C38). The same result is obtained using the Excel function MDURATION (cell C62), which requires inputs from the cells C53:C59 (except C57).

  Both the calculations yield a modified duration of 5.1914, which means that the price of the bond increases by EUR 5.1914 if the market interest rate increases by 1% (assuming that the bond price would be linearly dependent on the interest rate). However, this assumption is only an approximation, as the interest rate and the price of the bond are convex.
- Therefore, we use convexity in the next step. To calculate the convexity, the following net present value is calculated in the numerator:

  C44=(C7*D7*C8/(1+C56)^E7+D7*E7*D8/(1+C56)^F7+E7*F7*E8/(1+C56)^G7+F7*G7*F8/(1+C56)^H7+G7*H7*G8/(1+C56)^I7+H7*I7*(H8)/(1+C56)^J7)

  This present value is divided by the price of the bond (cell C45), which results in the modified reduction (cell C47). An Excel function for calculating the convexity is not available.

  The calculation of the convexity gives a value of 33.6791. The Convexity stands for the curvature of the curve of the bond price. It shows how a change in interest rates affects the duration. Since, we have a positive convexity here, we can conclude that this bond has low price sensitivity when interest rates rise and high price sensitivity when interest rates fall. This is an important additional information when we compare the bonds with the same or similar modified duration.

- The next step is to determine the Value at Risk, i.e., the loss if the interest rates increase. This can be done using an exact calculation or an approximate calculation.
- In the exact calculation, the difference between the NPV before and after the interest rate increase is simply calculated (cell C67). Since, the Value at Risk always has a positive value, we calculate the absolute value in the cell `C67=ABS(C22-C21)`. For example, if we assume that the interest rate does not increase by more than 1% at 99%, the maximum loss (Value at Risk) from this interest rate increase is EUR 51.59. The value of the exact calculation is the benchmark for the approximate calculations.
- The approximate calculation of the Value at Risk using the modified duration is given in the cell `C73=ABS(C21*C38*C15)`. This results in a Value at Risk of EUR 53.27. This shows that the risk is overestimated when using the modified duration.
- A better result is obtained by using the convexity in the cell `C74=ABS(-C38*C15*C21 +0.5*C47*C15^2*C21)`. Here, the Value at Risk is EUR 51.55 and thus comes very close to the exact value of EUR 51.59.
- You may wonder why we do a more detailed VaR calculation when we get the desired result with a simpler, more accurate calculation. Indeed, the approximate calculation does not make sense for a single bond. The situation is different if we have several securities with non-linear functions in our portfolio. Then, we have to rely on the approximate calculation.

Result (Figs. 10, 11, and 12)

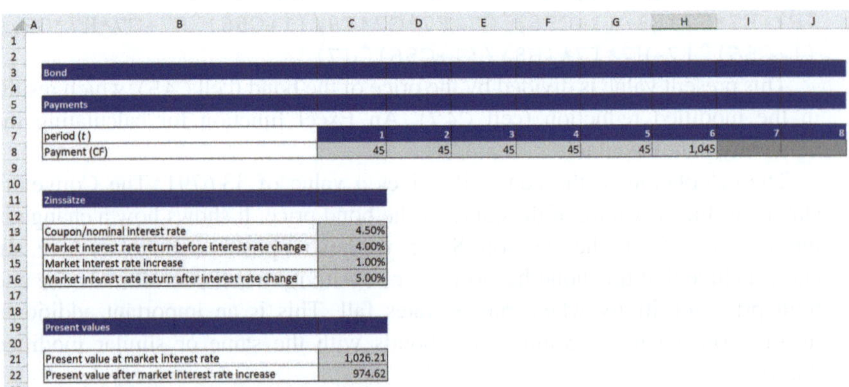

	B	C	D	E	F	G	H	I	J
3	Bond								
5	Payments								
7	period (t)	1	2	3	4	5	6	7	8
8	Payment (CF)	45	45	45	45	45	1,045		
11	Zinssätze								
13	Coupon/nominal interest rate	4.50%							
14	Market interest rate return before interest rate change	4.00%							
15	Market interest rate increase	1.00%							
16	Market interest rate return after interest rate change	5.00%							
19	Present values								
21	Present value at market interest rate	1,026.21							
22	Present value after market interest rate increase	974.62							

Fig. 10 Calculation of interest rates and cash values

	A	B	C
24			
25		**Duration**	
26			
27		$\sum_{t=1}^{T} t \cdot \frac{Z_t}{(1+r)^t}$	5,540.56
28		$\sum_{t=1}^{T} \frac{Z_t}{(1+r)^t}$	1,026.21
29			
30		Duration	5.3991
31			
32			
33		**Modified Duaration**	
34			
35		$\sum_{t=1}^{T} t \cdot \frac{Z_t}{(1+r)^{t+1}}$	5,327.47
36		$\sum_{t=1}^{T} \frac{Z_t}{(1+r)^t}$	1,026.21
37			
38		Modified Duration	5.1914
39		Control calculation: Modified Duration	5.1914
40			
41			
42		**Convexity**	
43			
44		$\sum_{t=1}^{T} t \cdot (t+1) \frac{Z_t}{(1+r)^{t+2}}$	34,561.81
45		$\sum_{t=1}^{T} \frac{Z_t}{(1+r)^t}$	1,026.21
46			
47		Convexität	33.6791
48			

Fig. 11 Calculation of duration, modified duration, and convexity

Literature and References to the Excel Tool

➲ Ernst, D. (2020): Corporate Risk Management, Course 2—Assignment 9.

Standard literature:

Hopkin, P. (2018): Fundamentals of Risk Management: Understanding, Evaluating and Implementing Effective Risk, 5. Edition, New York.

Hull, J.C. (2018): Risk Management and Financial Institutions, 5. Edition, Hoboken (New Jersey).

➲ See Excel file Case Study `Corporate Risk Management`, Excel worksheet `VaR non-linear functions (1)`

	A	B	C
50			
51		**Calculation with the Excel function DURATION and MDURATION**	
52			
53		Settlement date	31.12.2018
54		Due date	31.12.2024
55		Coupon/nominal interest rate	4.50%
56		Market interest rate return before interest rate change	4.00%
57		Market interest rate return after interest rate change	5.00%
58		Payment frequency	1
59		Base (to the day)	1
60			
61		Duration	5.3991
62		Mod. Duration	5.1914
63			
64			
65		**VaR: Exact calculation**	
66			
67		VaR (exact)	51.59
68			
69			
70		**VaR: approximate calculation**	
71			
72		VaR (Dur.)	53.27
73		VaR (mod. Dur.)	53.27
74		VaR (Con.)	51.55
75			

Fig. 12 Calculation of the duration and modified duration with Excel functions and calculation of the VaR

Assignment 10: Extreme Value Theory

Assignment

- Use the extreme value theory to calculate the Value at Risk and the conditional Value at Risk (the expected shortfall) for 1 day at a confidence level of 95% for the oil price from 31.12.2018 retroactively for the last 5 years.
- Perform a sensitivity analysis and determine the Value at Risk and the conditional Value at Risk (expected shortfall) at the following confidence levels: 99.0%, 99.1%, 99.2%, 99.3%, 99.4%, 99.5%, 99.6%, 99.7%, 99.8%, 99.9%.

Content

The concept of extreme events (e.g., disruptive technologies and financial market crisis), as described, for example, by *Taleb* in his "Black Swans," takes into account the problem that in real life there can be events that are not predictable (or difficult to predict), which can have serious economic consequences. This phenomenon is, generally, not taken into account in the traditional risk management. For the practice of risk management and for dealing with uncertainty in the future, it seems sensible to use known information from the past to draw conclusions about extreme values in the future. Techniques are needed here that avoid the use of a normal distribution and instead take into account that extreme events occur much more frequently than implied by a normal distribution. The methods required here are based on the fractal geometry of Mandelbrot and are essentially based on the so-called scalable distributions, such as the Pareto distribution. This gave rise to the extreme value theory, which we apply in the following:

Extreme value theory (EVT), which is also called Peaks-over-Threshold Method (PoT), goes beyond the expected shortfall, which is calculated on the basis of the available historical data. EVT is used for potentially catastrophic events that occur very rarely but produce extremely high sums of losses. EVT provides a scientific approach for predicting these rare events. It is a mathematical discipline that deals with the outliers, i.e., maximum and minimum values of samples.

Similar to the expected shortfall, EVT deals with the edges of distributions. It takes into account of empirical observations with extreme returns having a higher probability (so-called fat tails) and mean returns having a lower probability than described by a normal distribution. With the help of EVT, better VaR estimates can be obtained and VaR values for high confidence levels can be calculated accurately. In this respect, EVT is an extension of the concept of the expected shortfall, which measures in particular the impact of potentially extreme risks.

Extreme value theory is closely connected with the name *Gnedenko*. He was able to prove the main statement of EVT in 1943. It states that a large number of probability distributions have equal edges. This means that a broad class of distributions with increasing or decreasing returns r (depending on whether the maximum or minimum values of the sample are considered) strive against a generalized Pareto distribution.

EVT can explain both the left and right edges of a distribution. Here, we are dealing with the left edge of the distribution.

When using EVT, it should be noted that the Pareto distribution is only valid for "extreme" cases, i.e., for cases above a threshold to be determined by the user. The threshold should be set high enough that it allows the edge of the distribution to be actually examined, but it must also be set low enough that the number of data sets included in the maximum likelihood calculations is not too low, thus endangering the quality of the calculation.

Important Formulas

The (cumulative) generalized Pareto distribution is calculated with the following equation:

$$G_{\xi,\beta} = 1 - \left(1 + \xi \frac{y}{\beta}\right)^{-1/\xi} \tag{29}$$

The distribution has two parameters, namely ξ and β. The parameter ξ is a shape parameter that determines the severity of the edge of the distribution. The parameter β is a scaling parameter. y is the difference between r^{min} and $\widehat{r}$, where r^{min} is the minimum return in the distribution and $\widehat{r}$ is the critical threshold.

The parameters ξ und β can be estimated with the maximum likelihood:

$$\prod_{i=1}^{n_r} \frac{1}{\beta}\left(1 + \frac{\xi(-r_i + \widehat{r})}{\beta}\right)^{-\frac{1}{\xi}-1} \tag{30}$$

The maximization of the function of this function corresponds to the maximization of its logarithm:

$$\sum_{i}^{n_r} \ln\left[\frac{1}{\beta}\left(1 + \frac{\xi(-r_i + \widehat{r})}{\beta}\right)^{-\frac{1}{\xi}-1}\right] \tag{31}$$

Using the iterative search method (`SOLVER` in the Excel) we can now determine the parameters ξ and β, which maximize the last formula.

After determining the parameters ξ and β, the Value at Risk can be calculated. The formula is:

$$\mathrm{VaR} = \widehat{r} + \frac{\beta}{\xi}\left[\left(\frac{n}{n_{\widehat{r}}}(1-p)\right)^{-\xi} - 1\right] \tag{32}$$

n is the number of observations, $n_{\widehat{r}}$ is the number of values above the threshold. p is the confidence level.

The expected shortfall is calculated with the following formula:

$$\text{Expected Shortfall} = \frac{\mathrm{VaR} + \beta - \xi r}{1 - \xi} \tag{33}$$

Workbook: `Case Study Risk Management` PHARMA GROUP ➲ Worksheet: `Extreme Value Theory`

Calculation of the log-likelihood function in the Excel:

Excel example: `J12=IF(H12<$G$5; LN((1/$N$4)*((1+($N$5*(-$H12+$G$5)/$N$4)))^(-1/$N$5-1));0)`

Value at risk calculation in the Excel:

Excel example: `J5=-G5+(N4/N5)*(((J2/J3)*(1-J4))^(-N5)-1)`

Calculation of the expected shortfall in the Excel:

Excel example: `J6=(J5+$N$4+$G$5*$N$5)/(1-$N$5)`

Procedure

- First, the discrete return in the column `D` is calculated.
- The discrete daily returns are then sorted in ascending order of size in column `H` using the `SORT` function.
- Then, the confidence level p is set to determine the threshold value $\hat{r}$ (cell `G2`) and the alpha α is calculated (the cell `G3`) `G3=1-G2`.
- The next step is to determine the number of α-% smallest values. In our example, these are the 5% smallest values. With 1303 available yields, these are 65 values. Here, the function in the cell `G4=ROUNDDOWN((G3)*COUNT(D12:D1314);0)` is used.
- The return on the 65th lowest value is then determined. This is done for the 5% quantile with the function in the cell `G5=MIN(VLOOKUP(G4;G12:H1314;2;0);0)`. The calculated yield is -3.78%. The threshold value $\hat{r}$ is -3.78%.
- Column `J` contains the probability calculation for the maximum likelihood method. The Excel formula is: `J12=IF(H12<$G$5; LN((1/$N$4)*((1+($N$5*(-$H12+$G$5)/$N$4)))^(-1/$N$5-1));0)`
- To calculate the probability for the maximum likelihood method, we need the values of the parameters ξ and β in addition to the threshold value $\hat{r}$. We first obtain the initial values for the determination of ξ and β from the `Assumptions General`. We link this cell again with the cells `N4` and `N5`, which will contain the values of ξ and β after optimization with the `SOLVER`.
- In cell `J8`, the probabilities are summed `J8=SUM(J12:J1314)`, which are then maximized.
- For optimization using the `SOLVER`, the following values must be entered in the `SOLVER` (Fig. 13).

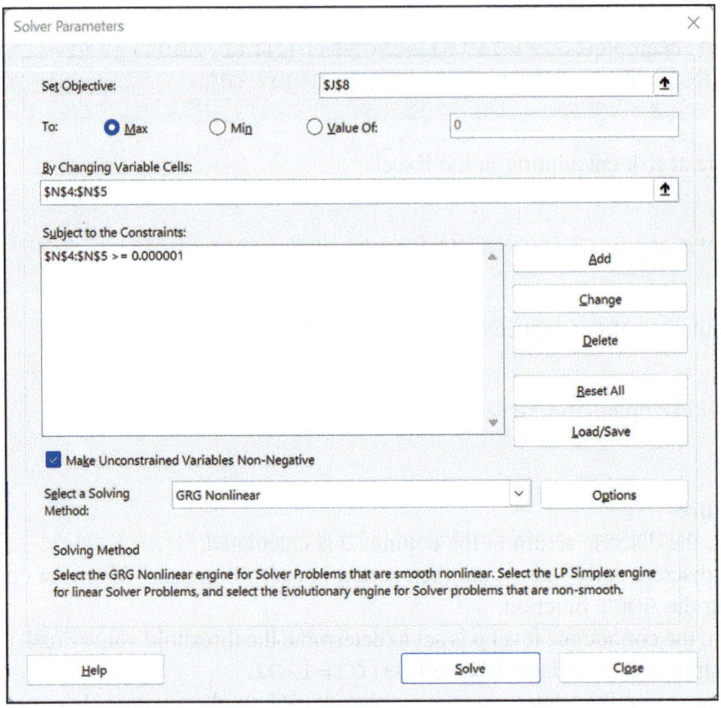

Fig. 13 Solver parameters for extreme value theory

- This results in values for β as 0.013932 and for ξ as 0.000010.
- In the next step, the Value at Risk can now be calculated with EVT. In addition to ξ and β, the number of observations n in the cell J2=COUNT(J12:J1314), the number of values $n_{\hat{r}}$, which are below the threshold in the cell J3=COUNTIF(J12:J1314; "<>0"), as well as the confidence level p and the threshold value $\hat{r}$ (cell G5) are required as variables. The VaR is calculated as follows J5=-G5+(N4/N5)*(((J2/J3)*(1-J4))^(-N5)-1). It amounts to 3.75% and, in contrast to the previous Value at Risk calculations, describes the exact value for the 5% quantile.
- Furthermore, the expected shortfall can be calculated with the extreme value theory
  J6 =(J5+$N$4+$G$5*$N$5)/(1-$N$5). It is 5.15%.
- With the help of a sensitivity analysis Data ➲ What-If-Analysis ➲ Data Table, the Value at Risk and the expected shortfall for different confidence levels can now be easily calculated.

Result (Fig. 14)

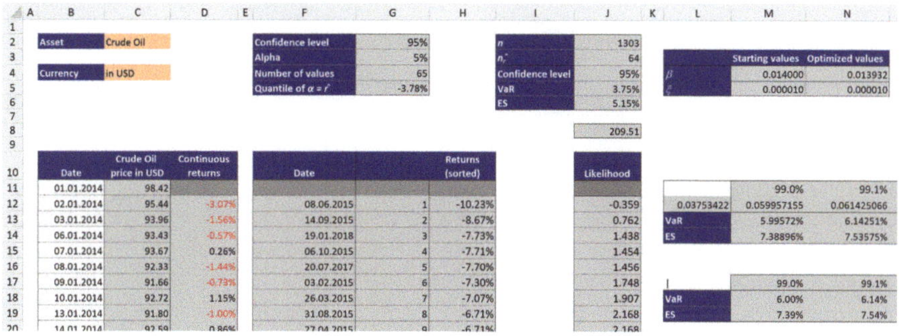

Fig. 14 Calculation of the VaR and the expected shortfall using the extreme value theory

Literature and References to the Excel Tool

Ernst, D. (2020): Corporate Risk Management, Course 2—Assignment 10.

Standard literature:

Hopkin, P. (2018): Fundamentals of Risk Management: Understanding, Evaluating and Implementing Effective Risk, 5. Edition, New York.

Hull, J.C. (2018): Risk Management and Financial Institutions, 5. Edition, Hoboken (New Jersey).

See Excel file Case Study `Corporate Risk Management`, Excel worksheet `Extreme Value Theory`

Assignment 11: Risk Measures in Comparison

Assignment

- Review the risk measure requirements listed in the following table.
- Complete the table.
- Justify your decision (Table 1).

Table 1 Risk measures in comparison

Requirement	Variance	Standard deviation	Semi-variance/semi-standard deviation	Absolute Value at Risk	Relative Value at Risk	Conditional Value at Risk	LPM_0	LPM_1	LPM_2
Easy interpretability									
Possibility of directly measuring the economic risk									
Use as target value for optimization problems									
Possibility of integrated risk measurement of different risk types									
Use for risk management of a portfolio									
Coherence									

Content

In general, the risk indicators or risk measures are used to quantify the risks and to take control measures based on them. The aim of risk quantification by means of certain risk measures is to identify and counteract the risks that threaten the company's existence.

To quantify the risk, a risk measure must be used that adequately reflects the level of risk. In general, the risk can be represented, as we have already done, in the form of a probability density or distribution function of a random variable. However, the illustration of risk in such a form is often not very meaningful and comprehensible for the non-experts, so that a condensation of the information is desirable.

As we have already seen with the risk indicators we used, the level of risk is determined by the extent to which the target is missed, i.e., the extent of the deviation from the expected value, and the probabilities to be attributed in each case. For this reason, the risk measures for risk management should allow statements to be made on the probability of occurrence on the one hand and the level of risk on the other. In addition, the selected risk measure should be easy to understand and interpret. It is, therefore, advisable to express risk in monetary units. The Value at Risk appears to meet these requirements.

Overall, a risk measure should serve to manage risks within the company. This means that the risk measure must be applied to control a large number of risks. It is, therefore, an aggregated risk measure that describes portfolio risk.

Artzner et al. have formulated the following four axioms that a measure of risk should fulfil within the framework of the risk control option. The risk measures that correspond to these characteristics are called *coherent risk measures*. Hull describes the axioms in a very understandable way (Hull, J. C. (2014), p. 227).

Axiom 1: *Monotony*: If a portfolio produces a result worse than another for each environmental condition, its measure of risk should be greater.

Axiom 2: *Translation invariance*: If the portfolio receives a cash of amount X, then the measure of risk should decrease by this amount X.

Axiom 3: *Homogeneity*: If the volume of the portfolio is changed by a factor of λ and the relative shares of the individual positions in the portfolio remain the same, then the measure of risk should also change by a factor of λ.

Axiom 4: *Sub-additivity*: The measure of risk for two merged portfolios should not be greater than the sum of the two measures of risk before the merger.

In summary, a risk measure should meet the following requirements:

- Easily interpretable
- Able to directly measure the economic risk
- Used as a target value for optimization problems

(continued)

- Able to integrate risk measurement of different risk types
- Used for risk management of a bank portfolio
- Coherent

Literature and References to the Excel Tool

Ernst, D. (2020): Corporate Risk Management, Course 2—Assignment 11.

Standard literature:

Hopkin, P. (2018): Fundamentals of Risk Management: Understanding, Evaluating and Implementing Effective Risk, 5. Edition, New York.

Hull, J.C. (2018): Risk Management and Financial Institutions, 5. Edition, Hoboken (New Jersey).

Course Unit 2: Determination of Portfolio Risks

Assignment 12: Variance-Covariance Method: Variance-Covariance Matrix and Portfolio Risk

Assignment

- Create a spreadsheet in which you copy the following values:
  - Oil prices from 31.12.2018 retroactively for the last 5 years.
  - 3 months Euro-EURIBOR from 31.12.2018 retroactively for the last 5 years.
  - EUR/USD exchange rates (USD price for 1 EUR) from 31.12.2018 retrospectively for the last 5 years.
- Calculate the daily continuous return for the given price data of the portfolio assets.
- Calculate the average returns for each portfolio asset.
- The following data is given for the portfolio assets:
  - Crude Oil: 100,000 barrels
  - Bond: US$2,500,000, maturity 1 year, Coupon rate: Euribor as of 31.12.2018 plus margin in the amount of 0.63%
  - Foreign currency holdings in US$: US$4,000,000
- Calculate the weights of the portfolio assets.
- Calculate the covariances for the assets in the portfolio and create a variance-covariance matrix.
- Calculate the portfolio return based on the historical returns of the portfolio assets and the portfolio weights.
- Calculate the portfolio risk (portfolio variance and portfolio standard deviation) based on the variance-covariance matrix and the portfolio weights.

D. Ernst, J. Häcker, *Corporate Risk Management*, Springer Texts in Business and Economics, https://doi.org/10.1007/978-3-031-53126-2_6

Content

The *variance–covariance method* is an analytical or parametric method. The term "parametric methods" is derived from the fact that parameters such as the standard deviation are used for calculation and the Value at Risk is not read as a quantile from an empirical distribution. Under the simplifying assumptions of the parametric approach, the expected value (mean value) and the risk (standard deviation) are sufficient to determine the Value at Risk.

In the variance–covariance method, a common normal distribution is assumed for the returns of various assets in the portfolio. Since the expected values of individual returns are known on the basis of the available empirical data, the portfolio return can be calculated without difficulty. It is also easy to determine the standard deviations of the individual returns and their covariances, so that portfolio risk can be calculated using the variance-covariance matrix. Using the portfolio return and the portfolio standard deviation, the Value at Risk, and the relative/conditional Value at Risk can now be derived for normally distributed data.

Important Formulas

The expected return of the portfolio is calculated as:

$$\boxed{\mu_P = \sum_{i=1}^{n} w_i \cdot \mu_i} \qquad (1)$$

μ_i = Expected value of r_i
w_i = Weight of the asset i

The formula for the covariance of a sample is:

$$\boxed{\mathrm{Cov}\left[r_{i,j}\right] = \sigma_{i,j} = \frac{1}{n-1}\sum_{t=1}^{n}(r_{i,t} - \mu_i)\cdot\left(r_{j,t} - \mu_j\right)} \qquad (2)$$

$\mathrm{Cov}[r_{i,j}] = \sigma_{i,j}$ = Covariance
i = Assets i
j = Assets j

The portfolio risk stated in terms of the portfolio variance is calculated with the following formula:

$$\sigma_P^2 = \sum_{i,j=1}^{n} w_i \cdot w_j \cdot \sigma_i \cdot \sigma_j \cdot \rho_{i,j} = \sum_{i,j=1}^{n} w_i \cdot w_j \cdot cov_{i,j} \qquad (3)$$

Workbook: `Case Study Risk Management PHARMA GROUP` ➲ Worksheet: `Variance Covariance Method`

The formula for calculating the average historical return, or the expected return, is:

> Excel example: `L19=AVERAGE(D19:D1321)`

The covariance in the variance-covariance matrix is calculated as follows:

> Excel example: `L23=COVARIANCE.S(G19:G1321;D19:D1321)`

The portfolio weights can be calculated as follows:

> Excel example: `L28=D5/D13`

The calculation for the portfolio return is:

> Excel example: `L33{=MMULT(L19:N19;L28:L30)}`

The portfolio variance is as follows:

> Excel example: `L34{=MMULT(MMULT(TRANSPOSE(L28:L30);L22:N24);L28:L30)}`

The portfolio standard deviation is the root of the portfolio variance:

> Excel example: `L35=SQRT(L34)`

Procedure

- Based on the price of oil (column C), the coupon (column F), and the EUR/USD exchange rate (column H), the continuous yields in columns D, G, and I are calculated as logarithmic yields. For the bond, the coupon was calculated in column F by adding the margin to the 3-month EURIBOR USD.
- The next step is to calculate the mean value of the historical returns, which serve as expected values in the cell L19=AVERAGE(D19:D1321).
- The basis for the calculation of the portfolio variance is the variance-covariance matrix, which is created as follows in the cell L22=COVARIANCE.S(D19:D1321;D19:D1321). It is calculated for a sample.
- Furthermore, the weights of the individual assets in the portfolio must be determined in the cell L28=D5/D13. Their sum is 100%. L31=SUM(L28:L30).
- Finally, the portfolio return, the portfolio variance, and the portfolio standard deviation are calculated. The portfolio return is calculated as L33{=MMULT(L19:N19;L28:L30)}, the portfolio variance as L34={MMULT(MMULT(TRANS(L28:L30);L22:N24);L28:L30)} and the portfolio standard deviation as L35=SQRT(L34).

Result (Fig. 1)

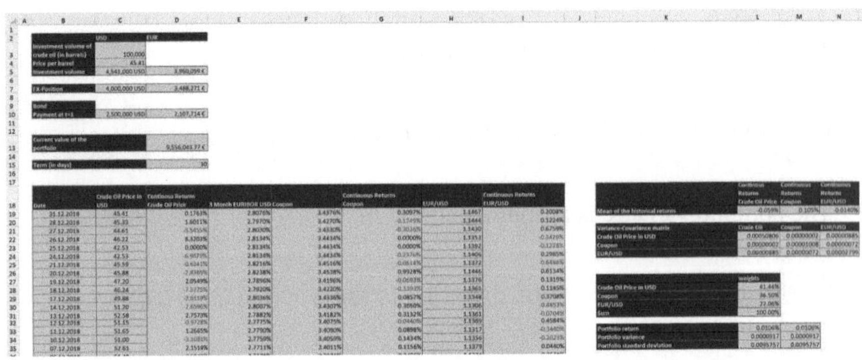

Fig. 1 Creation of the variance-covariance matrix and calculation of the portfolio risk using the variance-covariance method Literature and references to the Excel tool

Literature and References to the Excel Tool

Ernst, D. (2020): Corporate Risk Management, Course 2—Assignment 12.

Standard literature:

Hopkin, P. (2018): Fundamentals of Risk Management: Understanding, Evaluating and Implementing Effective Risk, 5. Edition, New York.

Hull, J.C. (2018): Risk Management and Financial Institutions, 5. Edition, Hoboken (New Jersey).

See Excel file Case Study `Corporate Risk Management`, Excel worksheet `Variance Covariance Method`

Assignment 13: Variance-Covariance Method: Calculation of Value at Risk and Conditional Value at Risk

Assignment

- Use the portfolio return and the portfolio standard deviation to generate a density function based on the assumption of a normal distribution.
- Calculate the portfolio Value at Risk for a confidence level of 5% and for a term of 1 day and 30 days based on the portfolio volume in EURO.
- Calculate the relative portfolio Value at Risk (deviation portfolio value-at-risk) for a term of 30 days based on the portfolio volume in EURO.
- Calculate the conditional portfolio Value at Risk (expected portfolio shortfall) for a term of 1 day and 30 days based on the portfolio volume in EURO.
- Enter the portfolio Value at Risk, the relative portfolio Value at Risk and the conditional portfolio value-at-risk in the density function.

Content

Since the variance-covariance method assumes a common normal distribution for the continuous returns of the various assets in the portfolio, the portfolio return and the portfolio standard deviation can be used to calculate the Value at Risk and the conditional Value at Risk for a normally distributed data. The procedure does not differ from the calculation of Value at Risk and the conditional Value at Risk for a continuous probability distribution.

Important Formulas

The formula of the Value at Risk is:

$$\boxed{\text{VaR}_{\propto}(r^c) = |\sqrt{N} \cdot \sigma_T \cdot z_\alpha - N \cdot \mu|} \tag{4}$$

The conditional Value at Risk is calculated as:

$$\boxed{\text{CVaR}_{(\alpha)} = \sqrt{N} \cdot \sigma_T . \frac{\varphi(Z_\alpha)}{\alpha} + N \cdot \mu_T} \tag{5}$$

Workbook: `Case Study Risk Management PHARMA GROUP` ➲ Worksheet: `Variance Covariance Method`

Calculation of the negative quantile of the distribution at a confidence level p assuming a normal distribution in the Excel:

Excel example: `L39=NORM.INV(L38;0;1)`

The 5% quantile is calculated in Excel with:

Excel example: `L42=L35*L39`

The Value at Risk is calculated in the Excel with:

Excel example: `L43=ABS(L35*L39-L33)`

For a given portfolio volume, the formula for the Value at Risk in monetary units for 1 day is:

Excel example: `L47=L43*L45`

Adding the time factor, the Value at Risk is 1.59% according to the Excel formula:

Excel example: `L52=ABS(L35*L39*L50-L33*L49)`

The formula for the Value at Risk in monetary units taking into account the time factor is:

Excel example: `L53=L45*L52`

For the conditional Value at Risk, the density is calculated in the Excel as follows:

Excel example: `L58=NORMDIST(L39;0;1;FALSE)`

Calculation of the conditional Value at Risk (expected shortfall) as a return with a continuous probability distribution in the Excel:

Excel example: `L59=L50*L35*L58/L38+L49*L33`

Calculation of the conditional Value at Risk (expected shortfall) in monetary units with a continuous probability distribution in the Excel:

Excel example: `L60=L45*L59`

Procedure

- When calculating the Value at Risk, the value of the standard normal distribution for the 5% quantile is first calculated in the cell `L39=NORM.INV(L38,0,1)`. This results in −1.645.
- The value of the 5% quantile is −1.58% `L42=L35*L39`.
- The Value at Risk for 1 day is 1.59% and is calculated with the Excel formula in the cell `L43=ABS(L35*L39-L33)`.
- With a portfolio volume of 9,556,044 €, the Value at Risk in monetary units for 1 day is 151,527 €. The Excel formula in the cell `L47=L43*L45` is used.
- Adding the time factor of 30 days, the Value at Risk is 8.95%, according to the Excel formula in the cell `M52=ABS(M35*M39*M50-M33*M49)`.
- The Value at Risk in monetary units is 854,809 €. The Excel formula `M53=M45*M52` is used.
- The Value at Risk indicates that in 95% of the cases, the loss from an investment in the portfolio does not exceed 854,809 € within the next 30 days.
- When calculating the conditional Value at Risk, the first step is to calculate the density for the 5% quantile `M58=NORMDIST(M39;0;1;FALSE)`. It amounts to 0.1031.

- The conditional Value at Risk is then calculated for 30 days in the cell `M59=M50*M35*M58/M38+M49*M33`. The conditional Value at Risk for 30 days is 11.14%.
- The conversion of the conditional Value at Risk into monetary units is affected by multiplying the conditional Value at Risk for 30 days by the portfolio volume in the cell `M60=M45*M59`. The conditional Value at Risk in monetary units is 1,064,239 €.
- The conditional Value at Risk states that in 5% of cases within 30 days the loss from the portfolio exceeds 854,809 €. The average loss then amounts to 1,064,239 €.

Result (Fig. 2)

Literature and References to the Excel Tool

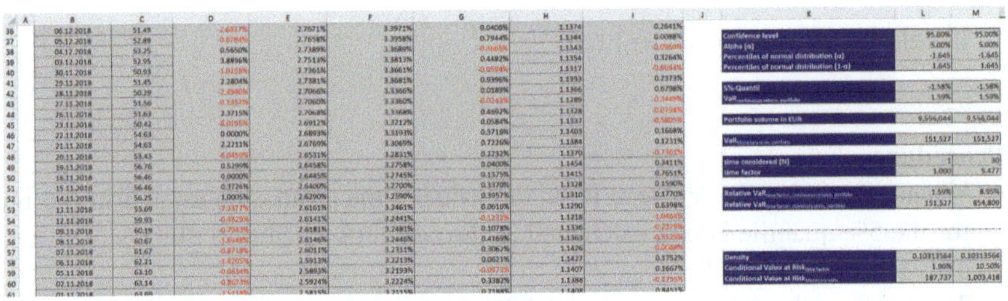

Fig. 2 Calculation of the Value at Risk and the conditional Value at Risk using the variance-covariance method

Ernst, D. (2020): Corporate Risk Management, Course 2—Assignment 13.

Standard literature:

Hopkin, P. (2018): Fundamentals of Risk Management: Understanding, Evaluating and Implementing Effective Risk, 5. Edition, New York.

Hull, J.C. (2018): Risk Management and Financial Institutions, 5. Edition, Hoboken (New Jersey).

See Excel file Case Study `Corporate Risk Management`, Excel worksheet `Variance Covariance Method`

Assignment 14: Historical Simulation

Assignment

- Create a spreadsheet in which you copy the following values.
  - Oil prices from 31.12.2018 retroactively for the last 5 years.
  - 3 months Euro-EURIBOR from 31.12.2018 retroactively for the last 5 years.
  - EUR/USD exchange rates (USD price for 1 EUR) from 31.12.2018 retrospectively for the last 5 years.
- Calculate the daily discrete returns for the given price data of the portfolio assets.
- Calculate the values of the simulated portfolio assets.
- Determine the simulated portfolio values.
- Calculate the profits/losses resulting from the difference between the simulated portfolio values and the current portfolio value.
- The following data is given for the portfolio assets:
  - Crude Oil: 100,0000 barrel
  - Bond: US$2,500,000, maturity 1 year, coupon: Euribor as of 31.12.2018 plus margin in the amount of 0.63%.
  - Foreign currency holdings in US$: US$4,000,000
- Calculate the portfolio Value at Risk for a confidence level of 5% and for a 1-day maturity based on the portfolio volume in EURO.
- Calculate the relative portfolio Value at Risk (deviation portfolio Value at Risk) for a term of 1 day based on the portfolio volume in EURO.
- Calculate the conditional portfolio Value at Risk (expected portfolio shortfall) for a 1-day term based on the portfolio volume in EURO.

Content

Historical simulation is often used for a portfolio for which the risk factors' historical data is available. Examples of such risk factors are share prices, exchange rates, commodity prices or interest rates.

The basic idea of a historical simulation is to calculate the changes in value for the past for each risk factor and apply these changes to the current value of the risk factor. Potential portfolio values are calculated for all the dates of the historical analysis. These are then used to calculate the gains and losses compared to the current portfolio value. The profits and losses determined are interpreted as a set of random values from which a frequency distribution can be determined. This, then, serves as a probability distribution for the future. The historical simulation, thus, implicitly assumes that the historical frequency distribution can be equated with the future probability distribution.

The historical simulation distinguishes between the factor approach and the portfolio approach. Under the *factor approach*, the Value at Risk is first calculated in isolation for the individual risk factors and then added together to form a portfolio Value at Risk. Under the *portfolio approach*, alternative portfolio values are generated by revaluing the portfolio with the historical values of the risk factors and the portfolio Value at Risk is then calculated.

A distinction can also be made between the difference method and the quotient method. The difference method uses the historical absolute changes (differences) between risk factors or portfolio values; *the quotient method* uses the historical relative changes in the form of the logarithmic changes.

In practice, the portfolio approach with the quotient method is mainly used.

Important Formulas

The result of the historical simulation is a discrete probability distribution for the profits and losses of the portfolio. For this reason, the already known formulas for Value at Risk and conditional Value at Risk are applied here for a discrete probability distribution.

Formally speaking, Value at Risk is the absolute value of the negative quantile of a distribution at the confidence level p. The formula for calculating the Value at Risk as a return for a discrete probability distribution is:

$$\boxed{\mathrm{VaR}_\alpha(P) = \mathrm{RVaR}_\alpha(P) = Q_\alpha(P)} \quad (6)$$

$\mathrm{VaR}_{\mathrm{Portfolio},\,\alpha}$ = Value at risk at the α quantile for the profit/loss of the portfolio
$\mathrm{RVaR}_\alpha\,(P)$ = Relative Value at Risk the α quantile for the profit/loss of the portfolio
P = Profit
$Q_\alpha = \alpha$ quantile

The formula for the calculation of the conditional Value at Risk as a return with a discrete probability distribution is:

$$\boxed{\mathrm{CVaR}_\alpha(P) = |E(P|P < \mathrm{Q}_\alpha(P))|} \tag{7}$$

CVaR_α = Conditional Value at Risk at confidence level p for the profit/loss of the portfolio

Workbook: `Case Study Risk Management PHARMA GROUP` ➲ Worksheet: `Historical Simulation.`

Calculation of the Value at Risk or relative Value at Risk as a return with a discrete probability distribution in the Excel:

Excel example: `W23=ABS(MIN(VLOOKUP(W21;S19:T1321;2;0);0))`

Calculation of the conditional Value at Risk (expected shortfall) as a return with a discrete probability distribution in the Excel:

Excel example: `W29=ABS(AVERAGE(T19:T83))`

Procedure

- Based on the oil prices (column `C`), the discrete returns are first calculated as a logarithmic returns (column `D`). Then the oil price is simulated, in which the possible oil prices for the next day are simulated with the given historical changes in the oil prices (column `E`). To do this, the potential future oil prices are calculated based on the current oil price by adjusting the current oil price for the historical changes in the oil prices `E19=$C$19*(1+D19)`. Finally, the simulated oil position in the portfolio is calculated by multiplying the potential future oil prices by the investment volume in barrels. The portfolio volume is then converted into euros at the current EUR/USD exchange rate (column `F`) `F19=(E19*$C$3)/$E$7`.
- The same procedure is used for the bond position (columns `G` to `K`) and the foreign currency position (columns `L` to `O`).
- After the individual positions of the portfolio have been calculated, they are added up to retrieve the portfolio value (column `P`) `P19=F19+K19+O19`.
- By deducting the current portfolio value from the potential portfolio values, the potential gains and losses are calculated (column `Q`) `Q19=P19-$D$13`.
- Based on the preliminary work, the price can now be sorted in ascending order `T19= SMALL($Q$19:$Q$1321;S19)`.

- Based on this list, the number of α-% smallest values is determined for a discrete probability distribution when determining the Value at Risk. In our example, these are the 5% smallest values. With 1303 available yields, this is 65.15 values. In this case, the result should be rounded down to 65 and the corresponding 65th value of the ordered list in the column `T` should be chosen. This is done for the 5% quantile with the function `W21=ROUNDDOWN((W19)*COUNT(T19:T1321);0)`. In our example, the 5% of the smallest values are 65 sets of values.
- The return on the 65th lowest value is then determined. This is done for the 5% quantile with the function W23=ABS(MIN(VLOOKUP(W21;S19:T1321;2;0);0)). The calculated Value at Risk is 145,530 €.
- Finally, the conditional Value at Risk can be calculated `W29=ABS(AVERAGE(T19:T83))`. It amounts to 201,827 €.

Result (Fig. 3)

Literature and References to the Excel Tool

Ernst, D. (2020): Corporate Risk Management, Course 2—Assignment 14.

Standard literature:

Hopkin, P. (2018): Fundamentals of Risk Management: Understanding, Evaluating and Implementing Effective Risk, 5. Edition, New York.

Hull, J.C. (2018): Risk Management and Financial Institutions, 5. Edition, Hoboken (New Jersey).

See Excel file Case Study `Corporate Risk Management`, Excel worksheet `Historical Simulation`

	N	O	P	Q	S	T
18	Simulated EUR/USD	Simulated Dollar Position in EUR	Simulated Portfolio Value	Profit and loss		Profit/Loss (sorted)
19	1.1490	3,481,274 €	9,555,819 €	- 225 €	1	- 400,993.31 €
20	1.1481	3,484,003 €	9,615,814 €	59,770 €	2	- 337,480.37 €
21	1.1545	3,464,771 €	9,394,815 €	- 161,229 €	3	- 302,460.08 €
22	1.1428	3,500,254 €	9,911,611 €	355,567 €	4	- 301,563.12 €
23	1.1453	3,492,558 €	9,560,331 €	4,287 €	5	- 292,740.11 €
24	1.1501	3,477,873 €	9,280,013 €	- 276,031 €	6	- 287,761.29 €
25	1.1393	3,510,970 €	9,553,755 €	- 2,289 €	7	- 286,493.36 €
26	1.1538	3,466,938 €	9,423,265 €	- 132,779 €	8	- 284,911.01 €
27	1.1482	3,483,671 €	9,633,709 €	77,665 €	9	- 276,031.03 €
28	1.1480	3,484,279 €	9,263,304 €	- 292,740 €	10	- 264,276.21 €
29	1.1510	3,475,360 €	9,440,978 €	- 115,066 €	11	- 259,833.17 €
30	1.1411	3,505,240 €	9,468,822 €	- 87,221 €	12	- 254,177.29 €
31	1.1459	3,490,727 €	9,668,992 €	112,948 €	13	- 250,292.25 €
32	1.1520	3,472,316 €	9,501,784 €	- 54,259 €	14	- 244,754.03 €
33	1.1428	3,500,292 €	9,618,473 €	62,429 €	15	- 242,139.82 €
34	1.1444	3,495,336 €	9,441,820 €	- 114,223 €	16	- 231,032.66 €
35	1.1472	3,486,738 €	9,640,568 €	84,525 €	17	- 227,853.14 €
36	1.1497	3,479,070 €	9,441,992 €	- 114,052 €	18	- 222,217.80 €

V	W
Confidence level	95%
Alpha	5%
Number of values	65
VaR = Relative VaR	145,530 €
Expected Shortfall	201,827 €

Fig. 3 Calculation of the Value at Risk and the conditional Value at Risk using the historical simulation

Assignment 15: Monte Carlo Simulation: Normally Distributed Risk Parameters

Assignment

- Create a spreadsheet in which you copy the following values:
  - Oil prices from 31.12.2018 retroactively for the last 5 years.
  - 3 months Euro-EURIBOR from 31.12.2018 retroactively for the last 5 years.
  - EUR/USD exchange rate (USD price for 1 EUR) of 31.12.2018 retroactively for the last 5 years.
- Calculate the daily continuous returns for the given price data of the portfolio assets.
- Calculate the continuous returns of each portfolio asset:
  - The mean value of the historical return, which corresponds to the expected value of the return.
  - The standard deviation of the sample.
- For Monte Carlo simulation, define a normal distribution with the calculated expected value and the standard deviation for each portfolio asset.
- Run the Monte Carlo simulation with 5000 drawings and determine the simulated portfolio values.
- Calculate the profits/losses resulting from the difference between the simulated portfolio values and the current portfolio value.
- Calculate the portfolio Value at Risk for a confidence level of 5% and for a 1-day maturity based on the portfolio volume in EURO.
- Calculate the relative portfolio Value at Risk (the deviation portfolio Value at Risk) for a term of 1 day based on the portfolio volume in EURO.
- Calculate the conditional portfolio Value at Risk (the expected portfolio shortfall) for a 1-day term based on the portfolio volume in EURO.

Content

Monte Carlo simulation is a stochastic method in which a frequently performed random experiments form the basis. Monte Carlo simulation attempts to solve the analytically unsolvable problems in a mathematical context numerically with the help of probability theory.

Monte Carlo simulation is a further development of the scenario analysis. Computer simulation is used to calculate and analyze a large representative number of the future scenarios (planning scenarios) in risk aggregation. In this way, a realistic range of future earnings and liquidity development is shown,

(continued)

i.e., the planning reliability or extent of a possible negative deviations from the plan.

The idea behind the Monte Carlo method is to determine the corresponding result or target figures for randomly selected parameters using the corresponding relationships (cause-effect network). The model used to determine the target values is usually deterministic in nature, which means that the target values are uniquely determined when the parameters are set. However, due to the random character of the parameters, the target variables are in principle random variables.

The law of large numbers serves as a justification for the Monte Carlo simulation. In its simplest form, this law states that the relative frequency of a random result usually stabilizes around the theoretical probability of a random result; and this is the case when the underlying random experiment is carried out repeatedly under the same conditions.

Monte Carlo simulation is not based on historical values like historical simulation, but on a stochastic variation of the different model parameters. This stochastic approach takes into account not only the individual risk positions and their influencing factors, but also the correlations to other risk positions.

Important Formulas

Similar to historical simulation, the formulas required here when using Monte Carlo simulation do not relate to the generation of the probability distribution of the target figure, but to the calculation of Value at Risk or conditional Value at Risk. The result of the Monte Carlo simulation is in turn a histogram or discrete probability distribution for the gains and losses of the portfolio. For this reason, the familiar formulas for the Value at Risk and conditional Value at Risk are used here with a continuous probability distribution.

Formally speaking, Value at Risk is the absolute value of the negative quantile of a distribution at the confidence level p. The formula for calculating Value at Risk as the return for a discrete probability distribution is:

$$\boxed{\mathrm{VaR}_\alpha(P) = \mathrm{RVaR}_\alpha(P) = Q_\alpha(P)} \tag{8}$$

VaR_α = Value at risk at the α quantile for the profit/loss of the portfolio
$\mathrm{RVaR}_\alpha\,(P)$ = Relative Value at Risk the α quantile for the profit/loss of the portfolio
P = Profit
Q_α = α quantile

The formula for the calculation of conditional Value at Risk as a return with a discrete probability distribution is:

$$\boxed{\mathrm{CVaR}_\alpha(P) = |E(P|P < \mathrm{Q}_\alpha(P))|} \tag{9}$$

$\mathrm{CVaR}_\alpha\, P =$ Conditional Value at Risk at confidence level p for the profit/loss of the portfolio

Workbook: `Case Study Risk Management PHARMA GROUP` ➲ Worksheet: `MCS Crystal Ball`

Calculation of the relative Value at Risk as a return with a discrete probability distribution in the Excel:

Excel example: `Y27=ABS(MIN(VLOOKUP(Y25;U23:V5022;2;0);0))`

Calculation of the conditional Value at Risk (expected shortfall) as a return with a discrete probability distribution in the Excel:

Excel example: `Y30=ABS(AVERAGE(V23:V272))`

Procedure

Preliminary remark: We will carry out the Monte Carlo simulation here using the software program Risk Kit as an example. Any other suitable software can, of course, be used. It should be noted here that the results of the Monte Carlo simulation are only displayed in the Excel file if the corresponding software is also installed on the computer.

- Based on oil prices (column `C`), the continuous returns are calculated (column `D`) first.
- Based on this we calculate the mean (cell D19) and the standard deviation (Cell D20). The normal distribution can be created using these two quantities.
- To create a normal distribution in Risk-Kit, we select:

the symbol for the normal distribution

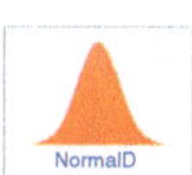

In the Function Type window, we select Single Random Number (Fig. 4)

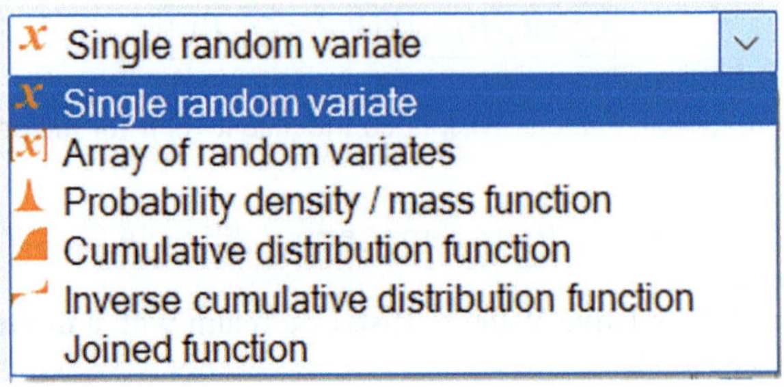

Fig. 4 Function arguments

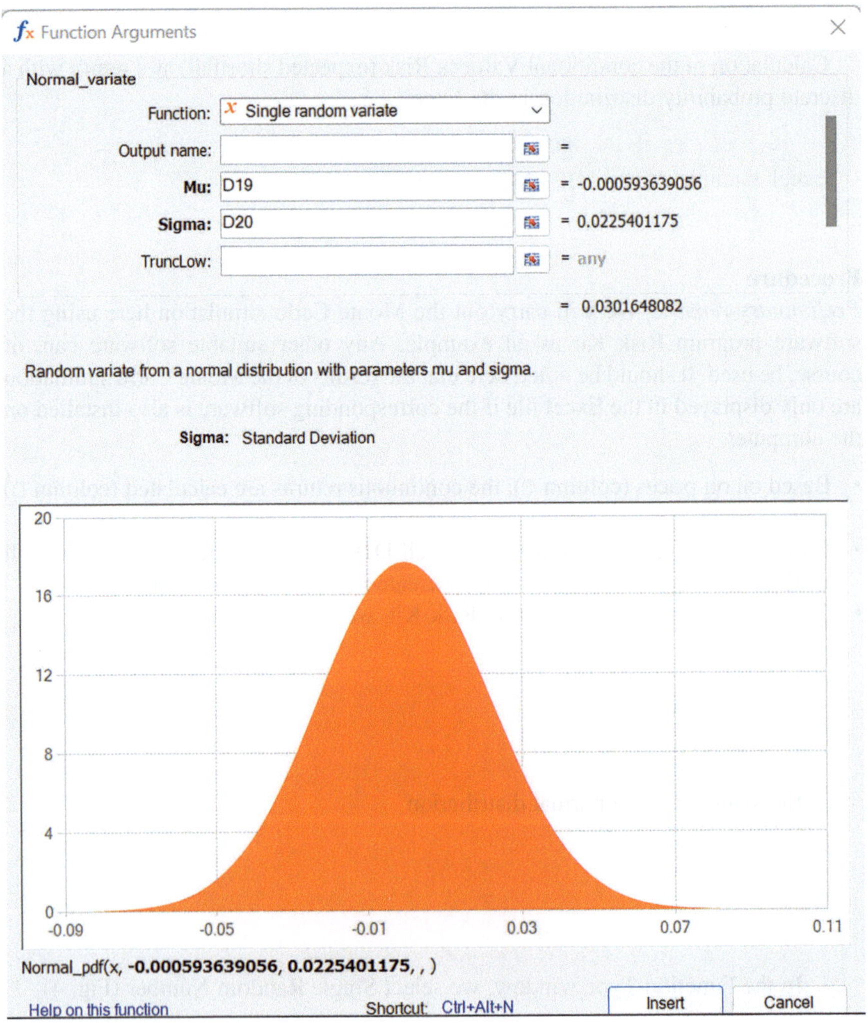

Fig. 4 (continued)

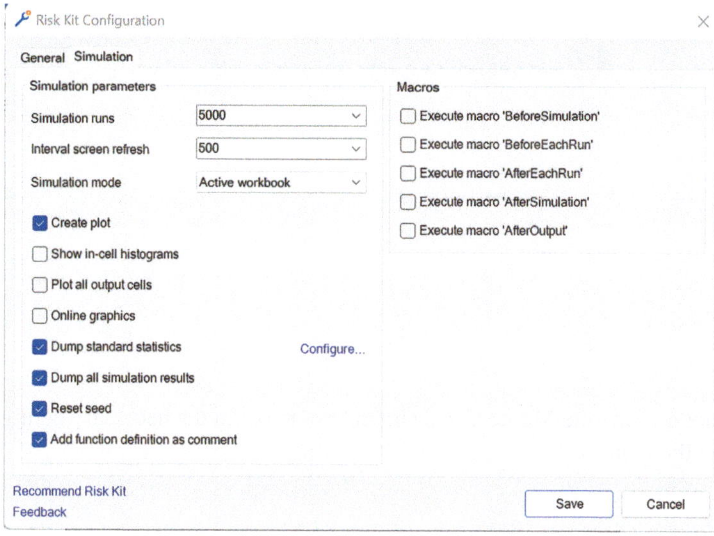

Fig. 5 Risk kit configuration

and can then enter the parameters for the normal distribution (Fig. 5).

- In the next step, a distribution function based on historical returns is created by calibration and entered as an assumption for the Monte Carlo simulation as an input parameter (cell `D18`). Cell D18 will now have the icon

defined as an input cell. Input cells for Monte Carlo simulation have a green background color in Risk Kit.
- The same procedure is used for the bond position (cell `I18`) and the foreign currency position (cell `M18`).
- After the input parameters for the Monte Carlo simulation have been entered, in the next step the portfolio value can be defined as the output parameters of the Monte Carlo simulation `P18=F18+K18+O18`. The Output cells for the Monte Carlo simulation will now have the icon

activated and have a dark yellow background color in Risk-Kit.

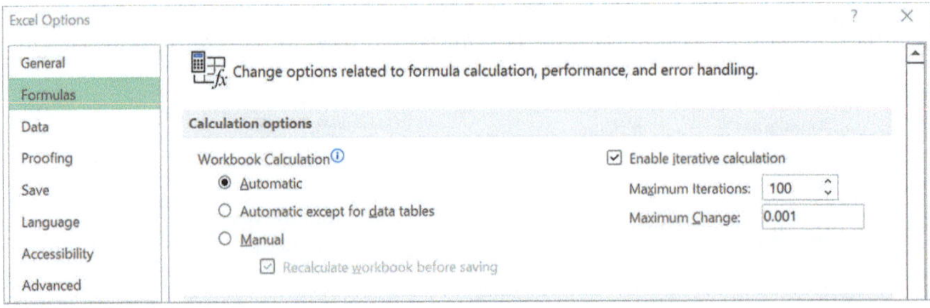

Fig. 6 Excel options

- Before running the Monte Carlo simulation, perform the necessary configuration about the icon

through (Fig. 5).

- After that, you can start the Monte Carlo simulation by clicking on the icon

to press.

- IMPORTANT: After running the Monte Carlo simulation, the calculation option for the workbook calculation is automatically reset to manual. Please switch this back to Automatic (Fig. 6).

- Column `R` lists the results of the Monte Carlo simulation (cells `R23:R5022`).
- By subtracting the current portfolio value from the potential portfolio values, the potential profits and losses are calculated (column `S`) `S23=R23-$P$20`.
- Based on the preliminary work, the profits can now be sorted in ascending order `V23 =SMALL($S$23:$S$5022;U23)`.
- On the basis of this list, the number of α-% smallest values is determined for a discrete probability distribution when determining the Value at Risk. In our example, the 5% smallest values are searched for. With the 5000 available returns, this means we are seeking 250 values. This is calculated with the function `Y25=ROUNDDOWN((Y23)*COUNT(V23:V5022);0))`.
- Then the return on the 250 smallest value is determined. This is done for the 5% quantile using the function `Y27=ABS(MIN(VLOOKUP(Y25;U23:V5022;2;0);0))`. The Value at Risk determined is 146,887 €.

- Finally, the conditional Value at Risk can be calculated `Y30=ABS(AVERAGE(V23:V272))`. It amounts to 182,612 €.

Result (Figs. 7 and 8)

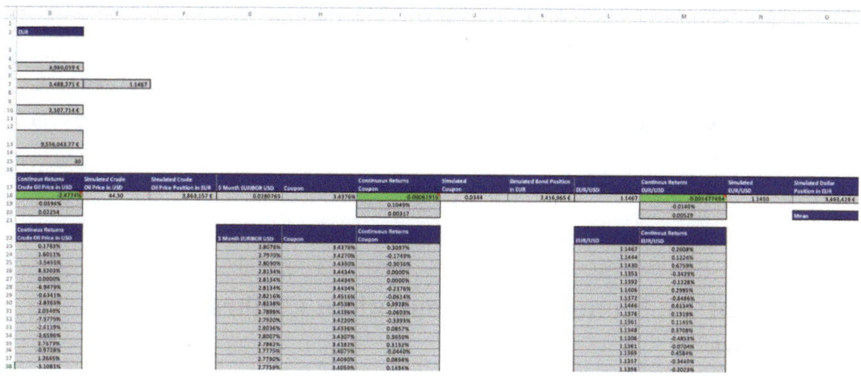

Fig. 7 Model conception of the Monte Carlo simulation based on a normal distribution

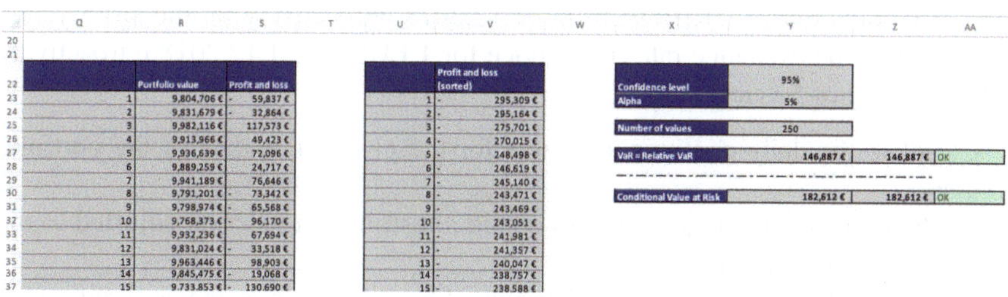

Fig. 8 Calculation of the Value at Risk and the conditional Value at Risk using the Monte Carlo simulation based on a normal distribution

Literature and References to the Excel Tool

Ernst, D. (2020): Corporate Risk Management, Course 2—Assignment 15.

Standard literature:

Hopkin, P. (2018): Fundamentals of Risk Management: Understanding, Evaluating and Implementing Effective Risk, 5. Edition, New York.

Hull, J.C. (2018): Risk Management and Financial Institutions, 5. Edition, Hoboken (New Jersey).

See Excel file Case Study `Corporate Risk Management`, Excel worksheet `MCS Crystal Ball`

Assignment 16: Monte Carlo Simulation: Calibrated Risk Parameters

Assignment

- Create a spreadsheet in which you copy the following values.
  - Oil prices from 31.12.2018 retroactively for the last 5 years.
  - 3 months Euro-EURIBOR from 31.12.2018 retroactively for the last 5 years.
  - EUR/USD exchange rate (USD price for 1 EUR) of 31.12.2018 retroactively for the last 5 years.
- Calculate the daily continuous returns for the given price data of the portfolio assets.
- Calibrate the continuous returns to obtain a distribution function adjusted to the given historical returns.
- Define the distribution function determined by calibration as the assumption for Monte Carlo simulation for each portfolio asset.
- Run the Monte Carlo simulation with the 5000 drawings and determine the simulated portfolio values.
- Calculate the profits/losses resulting from the difference between the simulated portfolio values and the current portfolio value.

- Calculate the portfolio Value at Risk for a confidence level of 5% and for a 1-day maturity based on the portfolio volume in EURO.
- Calculate the relative portfolio Value at Risk (the deviation portfolio Value at Risk) for a term of 1 day based on the portfolio volume in EURO.
- Calculate the conditional portfolio Value at Risk (the expected portfolio shortfall) for a 1-day term based on the portfolio volume in EURO.

Content

Monte Carlo simulation is carried out here in the same way as described above. The only difference is in the input parameters. In the Monte Carlo simulation with normally distributed input parameters, the risk parameters are normally distributed; in the Monte Carlo simulation with calibration, a distribution function that fits this data exactly is determined by a calibration on the basis of historical returns.

Important Formulas

Similar to the historical simulation and the Monte Carlo simulation, based on normally distributed risk parameters, the formulas required here when using a Monte Carlo simulation do not relate to the generation of the probability distribution of the target figure, but to the calculation of the Value at Risk, the relative Value at Risk and the conditional Value at Risk. The result of Monte Carlo simulation is in turn a discrete probability distribution for the gains and losses of the portfolio. For this reason, the familiar formulas for the Value at Risk and conditional Value at Risk are used here for a discrete probability distribution.

Formally speaking, the Value at Risk is the absolute value of the negative quantile of a distribution at the confidence level p. The formula for calculating the Value at Risk as return for a discrete probability distribution is:

$$\boxed{\mathrm{VaR}_\alpha(P) = \mathrm{RVaR}_\alpha(P) = \mid Q_\alpha(P) \mid} \tag{10}$$

$\mathrm{VaR}_{\mathrm{Portfolio}}$ = Value at risk at the α quantile for the profit/loss of the portfolio
RVaR_α (P) =Relative Value at Risk at the α quantile for the profit/loss of the portfolio
P = Profit
$Q_\alpha = \alpha$ quantile

The formula for the calculation of the conditional Value at Risk as return with a discrete probability distribution is:

$$\boxed{\mathrm{CVaR}_\alpha(P) = |E(P|P < \mathrm{VaR}_\alpha(P))|} \tag{11}$$

CVaR_α (P)= Conditional Value at Risk at confidence level p for the profit/loss of the portfolio

Workbook: `Case Study Risk Management PHARMA GROUP` ➲ Worksheet: `MCS Risk Kit`

Calculation of the negative quantile of the distribution at confidence level p in the Excel:

Excel example: `Y27=ABS(MIN(VLOOKUP(Y25;U23:V5022;2;0);0))`

Calculation of the conditional Value at Risk (expected shortfall) as a return with a discrete probability distribution in the Excel:

Excel example: `Y30 =ABS(AVERAGE(V23:V272))`

Procedure

Preliminary remark: We will carry out the Monte Carlo simulation here using the software program Risk Kit as an example. Any other, suitable software can, of course, be used. It should be noted here that the results of the Monte Carlo simulation are only displayed in the Excel file if the corresponding software is also installed on the computer.

- Based on the given oil prices (column `C`), continuous returns are calculated (column `D`) first.
- This will now be calibrated in the next step to get the best-fitting distribution function to be determined for this data series. To do this, we press the symbol

and choose Calibrate one-dimensional distributions.

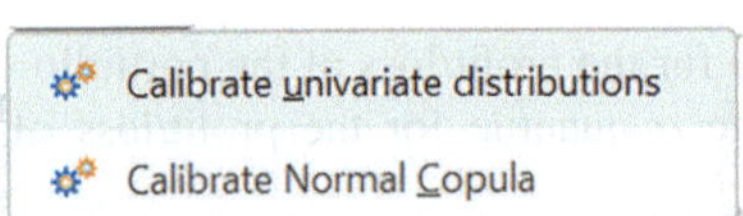

- A window will then open in which we can make the following settings. Then the calibration is carried out (Fig. 9).
- The following window will now appear in which you press OK (Fig. 10).
- The result of the calibration shows that the Laplace distribution is the distribution of the continuous and best reflects returns (Fig. 11).

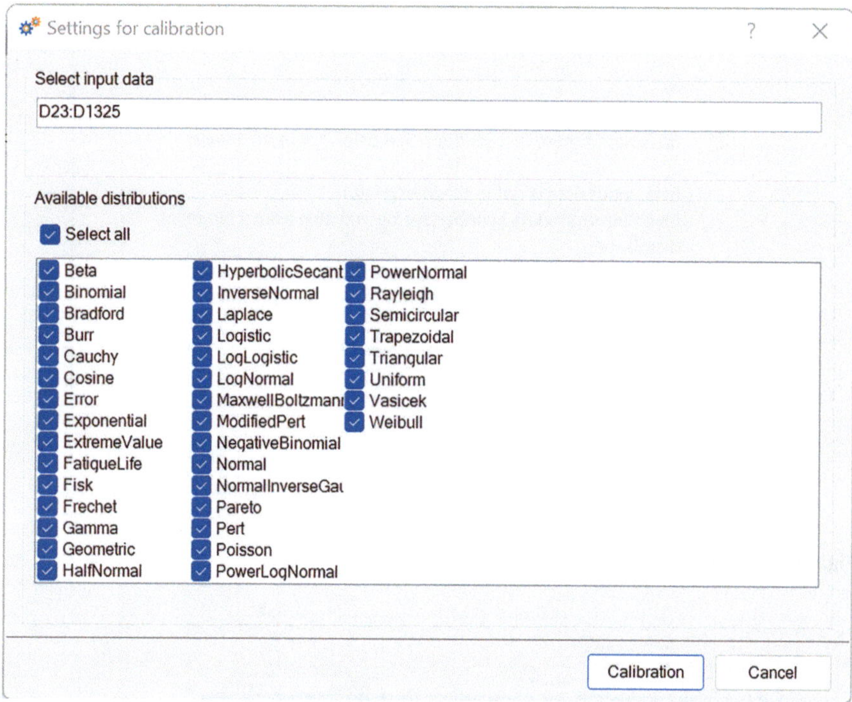

Fig. 9 Calibration

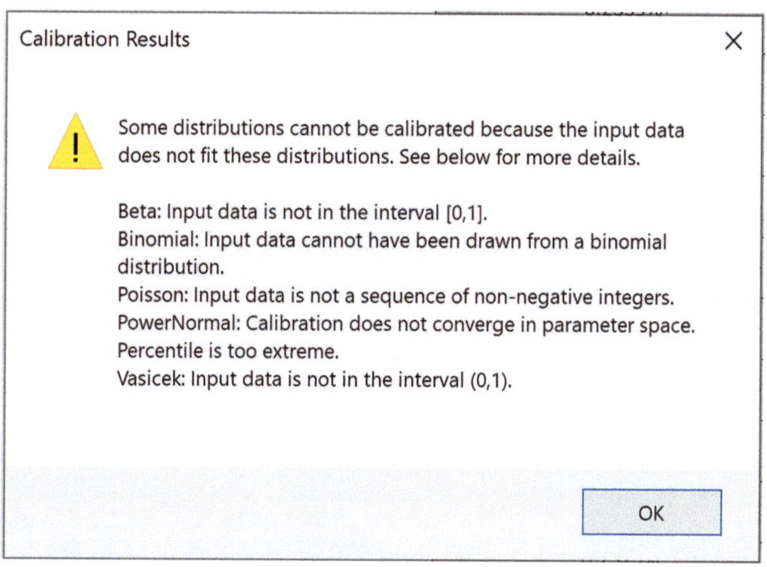

Fig. 10 Note on the calibration result

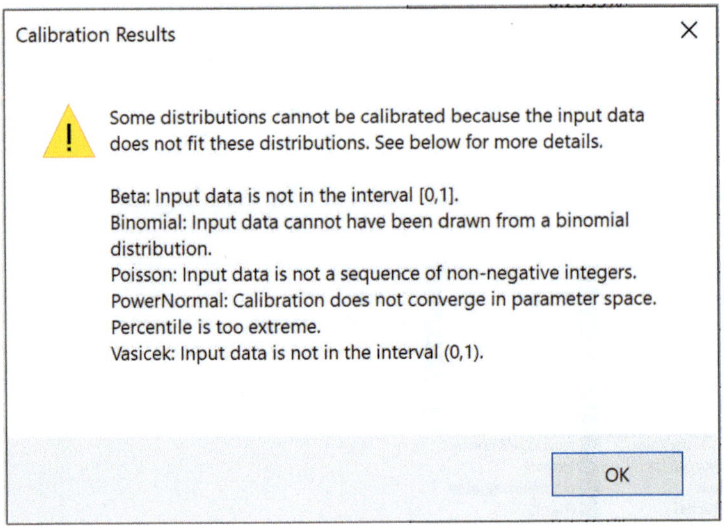

Fig. 11 Calibration result

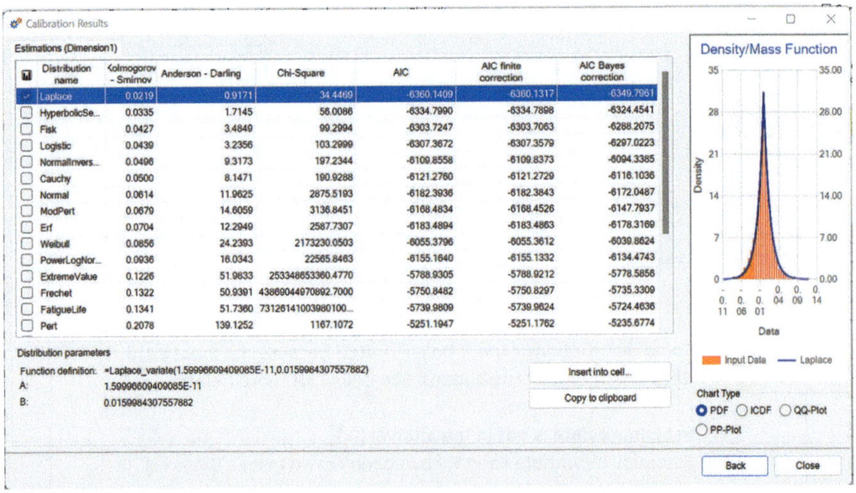

Fig. 11 (continued)

- Now select the Insert into cell... box and then enter cell D18.

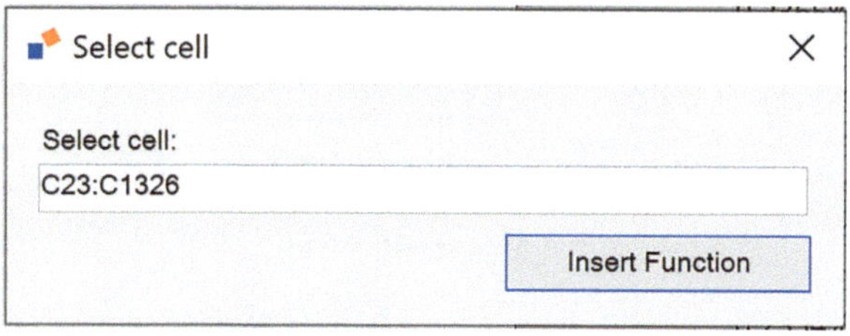

- The same procedure is used for the bond position (cell `I18`) and the foreign currency position (cell `M18`).
- After the input parameters for the Monte Carlo simulation have been entered, in the next step the portfolio value can be defined as the output parameters of the Monte Carlo simulation `P18=F18+K18+O18`. The Output cells for the Monte Carlo simulation will now have the icon

  activated and have a dark yellow background color in Risk-Kit.
- Before running the Monte Carlo simulation, perform the necessary configuration about the icon

  through (Fig. 12).

- After that you can start the Monte Carlo simulation by clicking on the icon

  to press.
- IMPORTANT: After running the Monte Carlo simulation, the calculation option for the workbook calculation is automatically reset to manual. Please switch this back to Automatic (Fig. 13).

- Column `R` lists the results of the Monte Carlo simulation (cells `R23:R5022`).
- By subtracting the current portfolio value from the potential portfolio values, potential profits and losses are calculated (column `S`) `S23=R23-$P$20`.
- Based on the preliminary work, the profits and losses can now be sorted in ascending order `V23=SMALL($S$23:$S$5022;U23)`.

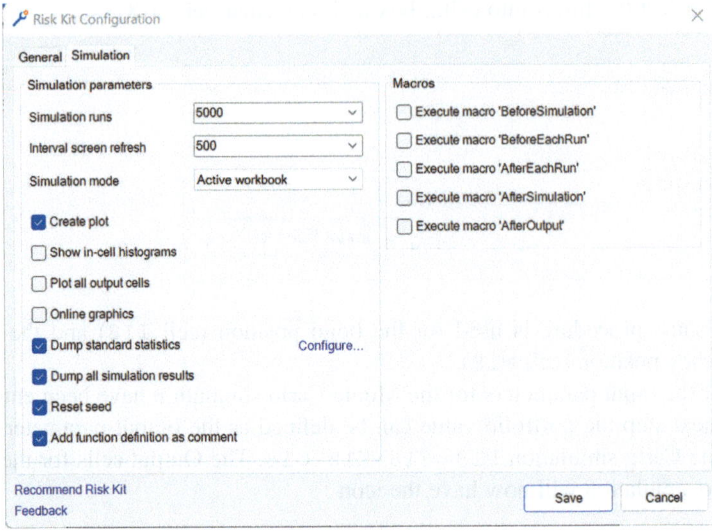

Fig. 12 Risk-kit configuration

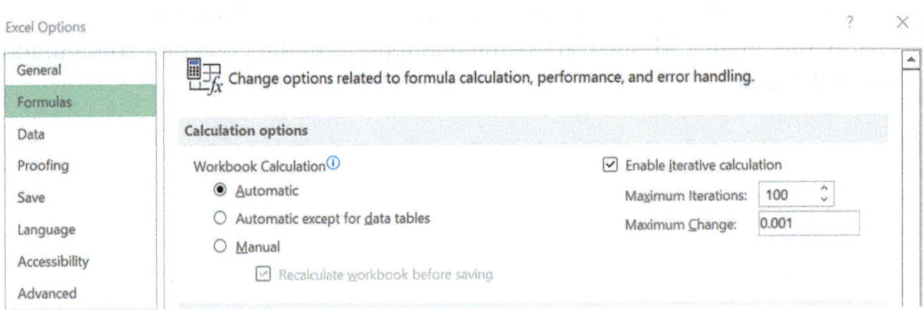

Fig. 13 Excel options

- On the basis of this list, the number of α-% smallest values is determined for a discrete probability distribution, when determining the Value at Risk. In our example, these are the 5% smallest values. With 5000 available returns, this means 250 values. The 5% quantile is statistically defined as the 250th worst value. In our VaR approach, we use the 250st value of the ordered list in Column `V`. This is calculated with the function `Y25=ROUNDDOWN((Y23)*COUNT (V23:V5022);0)`.
- The return on the 250st lowest value is then determined. This is done for the 5% quantile using the function `Y27=ABS(MIN(VLOOKUP(Y25;U23: V5022;2;0);0))`. The Value at Risk determined is 155,624 €.
- Finally, the conditional Value at Risk can be calculated `Y30=ABS(AVERAGE (V23:V272))`. It amounts to 217,828 €.

Result (Figs. 14 and 15)

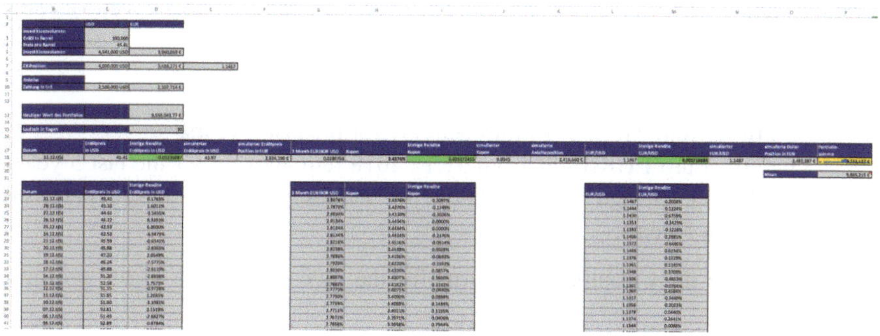

Fig. 14 Model conception of the Monte Carlo simulation based on a calibrated distribution

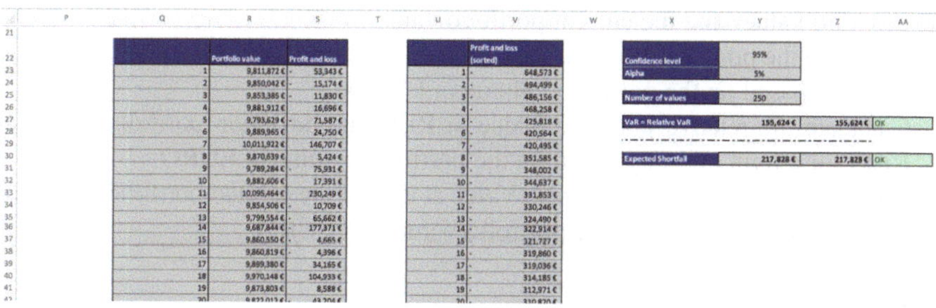

	Portfolio value	Profit and loss
1	9,811,872 €	- 53,343 €
2	9,850,042 €	- 15,174 €
3	9,853,385 €	- 11,830 €
4	9,881,912 €	16,696 €
5	9,793,629 €	- 71,587 €
6	9,889,965 €	24,750 €
7	10,011,922 €	146,707 €
8	9,870,639 €	5,424 €
9	9,789,284 €	- 75,931 €
10	9,882,606 €	17,391 €
11	10,095,464 €	230,249 €
12	9,854,506 €	- 10,709 €
13	9,799,554 €	- 65,662 €
14	9,687,844 €	- 177,371 €
15	9,860,550 €	- 4,665 €
16	9,860,819 €	- 4,396 €
17	9,899,380 €	34,165 €
18	9,970,148 €	104,933 €
19	9,873,803 €	8,588 €

	Profit and loss (sorted)
1	- 648,573 €
2	- 494,499 €
3	- 486,156 €
4	- 468,258 €
5	- 425,818 €
6	- 420,564 €
7	- 420,495 €
8	- 351,585 €
9	- 348,002 €
10	- 344,637 €
11	- 331,853 €
12	- 330,246 €
13	- 324,490 €
14	- 322,914 €
15	321,727 €
16	- 319,860 €
17	- 319,036 €
18	- 314,185 €
19	- 312,971 €

Confidence level	95%		
Alpha	5%		
Number of values	250		
VaR = Relative VaR	155,624 €	155,624 €	OK
Expected Shortfall	217,828 €	217,828 €	OK

Fig. 15 Calculation of the Value at Risk and the conditional Value at Risk using the Monte Carlo simulation based on a calibrated distribution

Literature and References to the Excel Tool

Ernst, D. (2020): Corporate Risk Management, Course 2—Assignment 16.

Standard literature:

Hopkin, P. (2018): Fundamentals of Risk Management: Understanding, Evaluating and Implementing Effective Risk, 5. Edition, New York.

Hull, J.C. (2018): Risk Management and Financial Institutions, 5. Edition, Hoboken (New Jersey).

See Excel file Case Study `Corporate Risk Management`, Excel worksheet `MCS Risk Kit`

Assignment 17: Monte Carlo Simulation Based on Copula-Functions

Course

- Create a spreadsheet in which you copy the following values:
  - Oil prices from 31.12.2018 retroactively for the last 5 years.
  - 3 months Euro-EURIBOR from 31.12.2018 retroactively for the last 5 years.
  - EUR/USD exchange rate (USD price for 1 EUR) of 31.12.2018 retroactively for the last 5 years.
- Calculate the daily continuous returns for the given price data of the portfolio assets.
- Calibrate the continuous returns with a Gaussian copula to obtain the correlations and the relevant distribution functions.
- Run the Monte Carlo simulation with 5000 drawings and determine the simulated portfolio values.
- Calculate the profits/losses resulting from the difference between the simulated portfolio values and the current portfolio value.
- Calculate the portfolio Value at Risk for a confidence level of 5% and for a 1-day maturity based on the portfolio volume in EURO.
- Calculate the relative portfolio Value at Risk (the deviation portfolio Value at Risk) for a term of 1 day based on the portfolio volume in EURO.
- Calculate the conditional portfolio Value at Risk (the expected portfolio shortfall) for a 1-day term based on the portfolio volume in EURO.

Content

Copulas are gaining popularity in the financial industry. Copulas can be used to model and estimate the marginal distributions separately from their dependency structure. They offer the possibility to use improved estimation methods for marginal distributions, since any assumptions about the correlation between the assets are already covered by the copula.

A *copula* is a function that can indicate a functional relationship between the marginal distribution functions of different random variables and their common probability distribution. The Copula functions can be used to combine random variables with arbitrary dependence structures to form a new common distribution function. There are different copula functions. The best known are the Gauss copula (normal copula), the student-t copula or the Clayton copula. In the following we use the Gaussian copula.

As a starting point, we imagine d risks, each of which is modelled as a random variable or represented by some historical data. This enables us to

(continued)

determine the univariate distribution function for each risk by calibration. However, this knowledge alone is not sufficient to determine the common distribution function of the risks, as we still lack all the information about the dependency between the individual variables. The Sklar theorem provides us with an important structure for determining their dependency. This states that any multivariate distribution function F can be split:

- Into a Copula C (a function which is itself a distribution function on the d-dimensional unit cube with uniform edges).
- The marginal distributions.

The marginal distributions are used as arguments in the copula and thus the common distribution is obtained.

The formal procedure is as follows:

- In a first step, the marginal distributions of the d risks are mapped to equally distributed [0.1] distributions. This is typically done using the corresponding quantiles of the marginal distributions (quantile mapping).
- In a second step, these are now mapped from independent confidence levels to correlated ones. The transformation rule and the associated dependency modelling is the actual Copula function.

The main property of a copula is to preserve the marginal distributions of the variables while defining a correlation structure between the variables. This is very convenient in many situations, as it separates the often very complicated analysis of the common distribution function into an analysis of the marginal distributions on the one hand and the copula on the other; in the latter, all information about the dependency is encoded. Conversely, a (mathematically correct) multivariate distribution function can always be constructed from any marginal distributions and any copula. This relationship is very popular in modelling and is therefore widely used, as it is a simple modular principle for linking individual risks. This modular principle also makes it easy to simulate a multivariate distribution. All that is needed is to simulate a sample according to the copula.

Despite the numerous possibilities of risk aggregation using Copula functions, we would like to point out the following critical observations. If the Copula function is chosen inappropriately for the problem at hand (and the typically pre-specified marginal distributions), a mathematically correct joint distribution can be generated, but this does not necessarily have to be economically sensible or problem-adequate. Often, the best known copulas (e.g., the Gaussian copula) are simply used without questioning their properties and effects on the current problem. It should also be noted that the choice of the copula function interferes with the result.

Important Formulas

Due to the complexity of the Copula functions, we cannot cover this exciting topic in detail here. However, we want to convey the basic idea of Copulas and its importance for risk management. For this purpose, we choose the Gaussian Copula.

The normal or Gaussian copula is defined using the distribution function of the normal distribution *F(·)*. So is:

$$\boxed{C(u_1,u_2)=F_2\left(F^{-1}(u_1),F^{-1}(u_2),\rho\right)} \tag{12}$$

a copula, where $F_2(\cdot,\ \cdot,\rho)$ is the bivariate distribution function of two standard normally distributed random variables with the correlation coefficient ρ.

If you generate points that are distributed according to the normal copula with the parameter $\rho = 0.5$, there is already a slight concentration of these along the bisector.

The Gauss Copula cannot be modelled easily with Excel. Hull shows the function of a Gaussian copula with a bivariate distribution function quite meticulously in his book. Due to the complexity of the calculation, a macro is already used here, which John Hull proposes for the calculation of a cumulative bivariate normal distribution (http://www-2.rotman.utoronto.ca/~hull/).

The computations of the multivariate distribution functions and of the copula functions that go beyond the Gaussian copula are even more complex. Special software tools are required to solve these problems. For the following calculations, we use Risk Kit, which allows a calibration of the multivariate distribution functions using a Gaussian copula.

Procedure

Workbook: `Case Study Risk Management PHARMA GROUP` ➲ Worksheet: `Copulas`

- *Preliminary remark:* We carry out the calculation with the help of Copulas with the software program Risk Kit. Any other suitable software can, of course, be used, but it must include a calibration function. It should be noted here that the results of the Monte Carlo simulation are only displayed if the corresponding software is also installed on the computer.
- Based on the oil prices (column `C`—cells `C23:C1326`), the coupon of the bond (column `D`—cells `D23:D1326`), and the EUR/USD exchange rate (column `E`—cells `E23:E1326`), the continuous returns of the individual positions are calculated first (columns `F:H`—cells `F23:H1326`).
- It is important for the subsequent process that the returns of the individual assets are directly adjacent to each other in a contiguous cell area (here `F23:H13256`).
- In the next step, the copula and the marginal distributions are estimated.
- To do this, select the cell range (`F23:H1325`).
- Now go to the Risk Kit on ➲ `Calibration` ➲ `Calibrate Normal Copula` (Fig. 16)

- Under `Available Distributions` choose `Select all` (Fig. 17).

Fig. 16 Calibrate Normal Copula

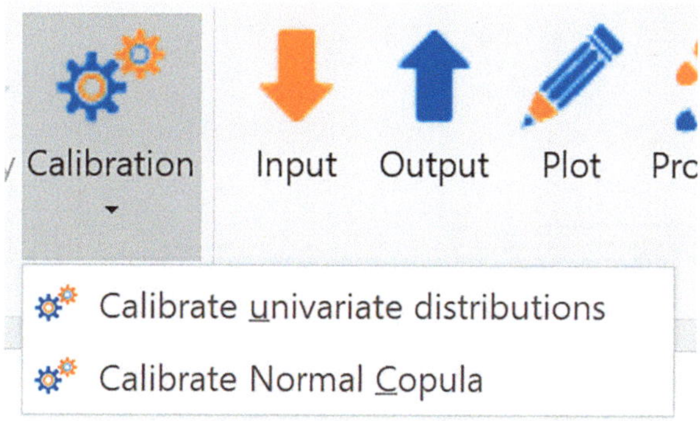

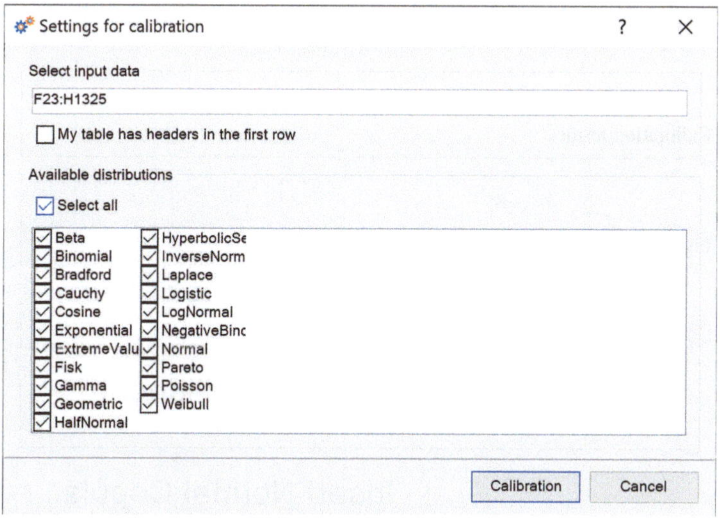

Fig. 17 Setting for calibration

- In the next step the result of the calibration is obtained (Fig. 18).
- Click `Insert Normal Copula` and specify a cell range large enough for the output (Fig. 19).
- The information transferred by the Risk Kit contains the correlations (cells `G2:J5`) and the calculated distribution functions with the parameters (cells `G7:J11`) that describe the distribution. These values are used as the basis for estimating the random numbers (cells `H14:J14`) used in the Monte Carlo simulation (Fig. 20).

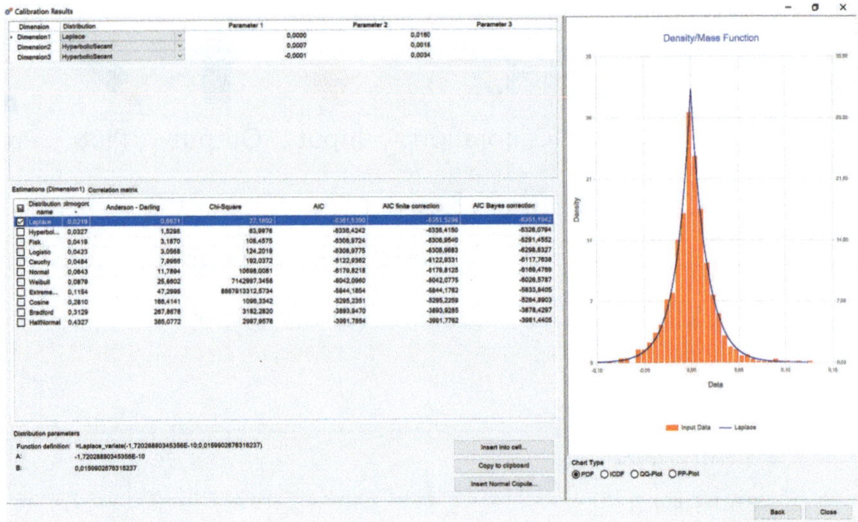

Fig. 18 Calibration results

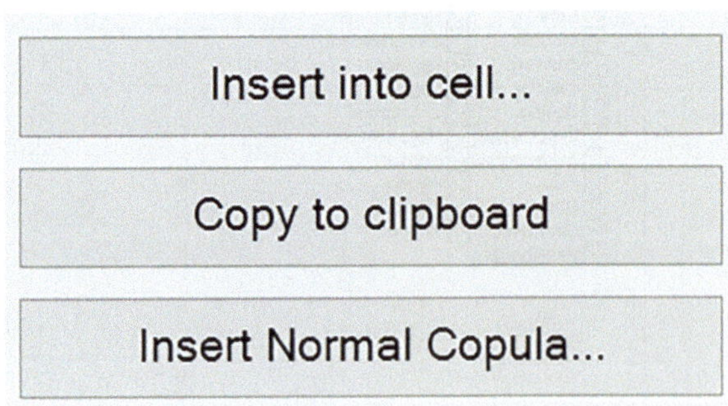

Fig. 19 Insert Normal Copula

- The three random numbers of the three assets (cells `H14:J14`) serve as the input cells for the Monte Carlo simulation. They have a green background color in Risk Kit.
- Once the Monte Carlo simulation input parameters have been entered, the next step is to define the portfolio value as the Monte Carlo simulation output parameters `P18=F18+K18+O18`. The Monte Carlo simulation output cells in Risk Kit have a dark yellow background color.

Correlation	log. Rendite Crude Oil in USD	log. Rendite Kupon	log. Rendite EUR/USD
log. Rendite Crude Oil in USD	1	-0.00445810	0.06982086
log. Rendite Kupon	-0.00445810	1	0.02655303
log. Rendite EUR/USD	0.06982086	0.02655303	1

	log. returns Crude Oil in USD	log. returns Cupon	log. returns EUR/USD
Distribuation	Laplace	HyperbolicSecant	HyperbolicSecant
Parameter 1	0	0.00073676	-0.00012611
Parameter 2	0.01599904	0	0.00336752
Parameter 3			

	log. returns Crude Oil in USD	log. returns Cupon	log. returns EUR/USD
Random numbers	0.015854635	0.001871103	0.00210589

Fig. 20 Correlations and the determined distribution functions with the parameters

- The results of the Monte Carlo simulation are listed in the columns `S` and `T` (cells `S23:T5022`). Column `T` shows the portfolio values, column `T` the profits and losses.
- Based on the preliminary work, the profits and losses can now be sorted in ascending order `W23 =SMALL($T$23:$Z$5022;V23)`.
- On the basis of this list, the number of α-% smallest values is determined for a discrete probability distribution when determining the Value at Risk. In our example, these are the 5% smallest values. With 5000 available returns, this means 250 values. The 5% quantile is statistically defined as the 250th worst value. In our VaR approach, we use the 250th value of the ordered list in column `X`. This is calculated with the function `Z25=ROUNDDOWN(Z23*COUNT(W23:W5022);0)`.
- The loss of the 250th lowest value is then determined. This is done for the 5% quantile using the function `Z27=ABS(MIN(VLOOKUP(Z25;V23:W5022;2;0);0))`. The Value at Risk determined is 144,414 €.
- Finally, the conditional Value at Risk (expected shortfall) can be calculated `Z30=ABS(AVERAGE(W23:W272))`. It amounts to 203,476 € (Figs. 21 and 22).

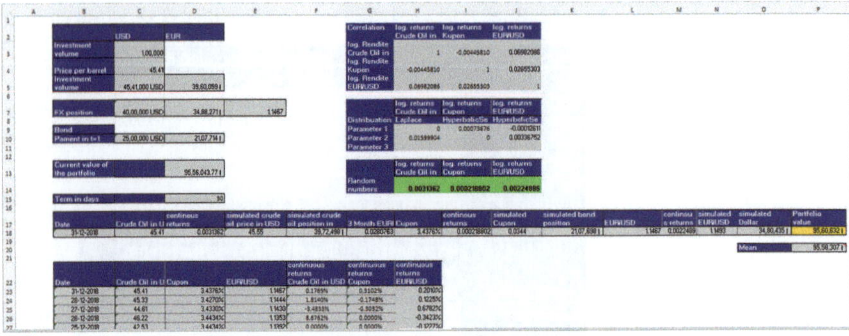

Fig. 21 Model conception of the Monte Carlo simulation based on a copula

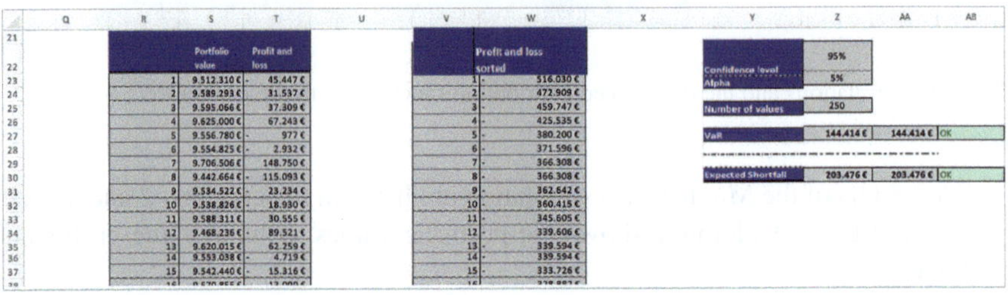

Fig. 22 Calculation of the Value at Risk and the conditional Value at Risk using the Monte Carlo simulation based on a copula

Literature and References to the Excel Tool

Ernst, D. (2020): Corporate Risk Management, Course 2—Assignment 17.

Standard literature:

Hopkin, P. (2018): Fundamentals of Risk Management: Understanding, Evaluating and Implementing Effective Risk, 5. Edition, New York.

Hull, J.C. (2018): Risk Management and Financial Institutions, 5. Edition, Hoboken (New Jersey).

See Excel file Case Study `Corporate Risk Management`, Excel worksheet `Copula`

Course Unit 3: Hedging of Hedgeable Risks and Modelling of Non-hedgeable Risks

Content

The hedging of entrepreneurial risks is one of the most important tasks of corporate risk management. The starting point for risk hedging is a systematic inventory of existing risks in the company. The following risks in the company can be distinguished:

- Strategic risks (for example, a product of the company is replaced by a technological innovation)
- Market risks (for example, market entry of a new competitor and price fluctuations on the procurement or input market)
- Financial market risks (for example, currency or interest rate risks)
- Legal and political risks (for example, changes in tax legislation, liability risks, and cyber risks)
- Risks of personnel management (for example, lack of employee motivation and poor working atmosphere)
- Performance risks (for example, technical risks in production or IT)

If the risk positions in the company are known, suitable measures can be taken to hedge them. Measures for this are:

- Risk avoidance (for example, not serving certain customers or markets)
- Risk reduction by reducing the probability of occurrence or the amount of damage (for example, diversification of the customer and/or supplier portfolio)
- Transferring risks to third parties (for example, taking out insurance, hedging interest rate, and currency or commodity risks)
- Creation of adequate risk coverage potential (for example, increase in equity position)

(continued)

D. Ernst, J. Häcker, *Corporate Risk Management*, Springer Texts in Business and Economics, https://doi.org/10.1007/978-3-031-53126-2_7

In the following, we distinguish between the following forward transactions:

- Risks that can be hedged (hedgeable risks)
  - Hedged risks
  - Non-hedged risks
- Risks that cannot be hedged (non-hedgeable risks)

The Risks that can be hedged can be hedged, for example, by insurance or hedging. We deal with hedging in more detail in Assignments 1–3. However, a company can also deliberately refrain from hedging even hedgeable risks. The reasons for this may be the associated costs or the conviction with which the risks can be correctly assessed and borne. In these cases, the risks are referred to as non-hedged risks. In addition to non-hedged risks, there are also non-hedgeable risks. These include all risks related to the entrepreneurial activities of the company and constitute entrepreneurship. Examples of such non-hedgeable risks are the losses of a customer, insolvency of a supplier, bad investment in a new production facility, new competitors and new technologies substituting previous products, legal risks, reputational risks, etc. The non-hedged and non-hedgeable risks together make up the business risks to which the company is exposed. These must be considered in a business plan. Assignments 4–6 show how these risks are integrated into the business plan and how the risk aggregation and risk analysis are performed using Monte Carlo simulation.

The following figure provides an overview of various forward transactions and the possible hedging instruments. These are also called derivatives. The word derivative comes from the Latin word "derivare" (to derive) and refers to a financial instrument that is derived from another financial instrument (Fig. 1).

Forward transactions are used in risk management to avoid and reduce risk. Their purpose is to hedge the performance of an underlying object, also known as an underlying. The underlying can be a commodity price, an exchange rate, an interest rate or even a share price. Depending on their trading platforms, a distinction can be made between the exchange-traded and over-the-counter futures. Over-the-counter (OTC) contracts are traded directly between the buyer and the seller. In the case of exchange-listed contracts, also known as contracts, trading takes place via the stock exchange. All the forward transactions involve at least two contracting parties. The position of the buyer, who is hoping for an increase in value (the long position), and the position of the seller, who is speculating on a fall in price (the short position).

There are a large number of differently structured forward transactions. Depending on the structure, term and underlying, these can be very complex. They range from simple derivatives, including plain vanilla derivatives, to

(continued)

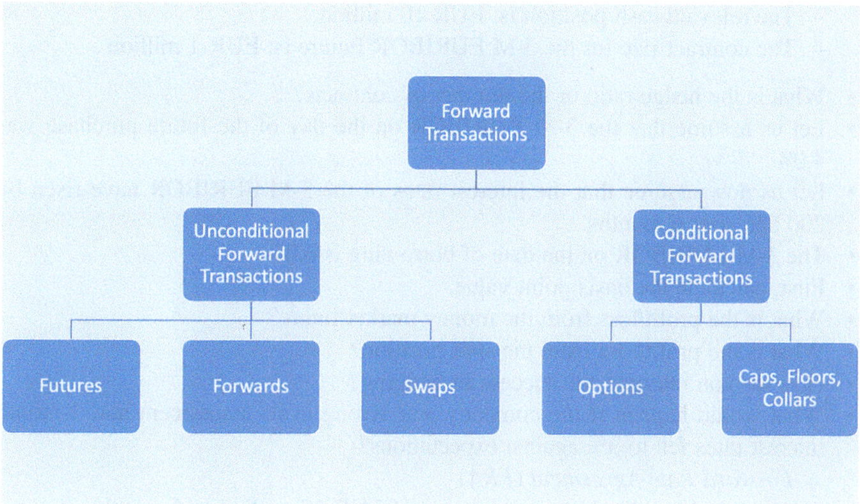

Fig. 1 Forward transactions

complex exotic derivatives. The flexible design of derivatives has great advantages for the companies' hedging activities. However, they also carry great risks, especially when the workings of a hedging instrument are not understood. In such cases, a hedging instrument can turn into a speculation that entails going-concern risks.

We, therefore, refer explicitly to the book "Derivatives: Options and Futures Step by Step," which discusses strategies with derivatives in detail.

In corporate risk management, simple derivatives are of particular importance, as they account for the majority of daily hedging activities. In the following, we would like to take interest rate risks as an example with

- A Future/Forward
- A swap
- Hedge with a cap

Assignment 18: Hedging Interest Rate Risk Using Money Market Futures and Forward Rate Agreements (FRAs)

Task

Money Market Futures

- A company has a liquidity requirement of EUR 10 million in 6 months.
- The company then needs this liquidity for 3 months.

- The relevant cash position is: EUR 10 million.
- The contract size for the 3-M EURIBOR Future is: EUR 1 million

- What is the hedge ratio or the number of contracts?
- Let us assume that the 3-M EURIBOR on the day of the future purchase was 4.0%.
- Let us now assume that the interest rates of the 3-M-EURIBOR have risen by 200 bps after 6 months.
- The 3-M-EURIBOR on the date of borrowing is 6.0%.
- First, calculate the basis point value.
- What is the profit/loss from the money market future?
- What is the profit/loss from the spot position?
- How do you interpret the success of hedging?
- What would happen if the company was wrong in its assessment and 3-month interest rates fell to 3% against expectations?

Forward Rate Agreement (FRA)

- A company has a liquidity requirement of EUR 10 million in 6 months.
- The company then needs this liquidity for 3 months.
- Based on its rating, the company would currently have to pay the 3-M-EURIBOR plus a risk premium of 1% for a 3-month loan.
- Since the company expects interest rates to rise, it buys an FRA at the FRA interest rate of 4.0%.
- The reference interest rate is set at 3 M EURIBOR + 1%.
- Two days before the start of the FRA period, the 3-M EURIBOR is 5.0%. This fixes the reference interest rate at 6.0%.
- What is the gain/loss from the FRA?
- What is the gain/loss on the cash position when borrowed?
- How do you interpret the success of hedging?
- What is the actual payout amount at the beginning of the forward?
- You want to determine the price of the FRA, that is, the FRA interest rate.
- The following initial values are given:

  - Total running time in days: 270
  - Term of the previous period in days: 180
  - Duration of the FRA period in days: 90
  - r_{t+T} (interest rate for the total term) 5.5%
  - r_t (interest rate for the previous period) 5.0%

- What is the price of the FRA, i.e., the FRA interest rate r_{FRA}?
- You want to determine the price of the FRA, i.e., the FRA interest rate, taking into account the bid and ask prices.
- The following initial values are given:

  - T_{t+T} (total running time): 1.00
  - T_t (term of the previous period): 0.67
  - $T_{FRA} = T$(FRA period maturity): 0.33
  - $r_{t+T}{}^b$ (bid price) 5.45%

- r_{t+T}^{o} (ask price) 5.55%
- r_t^{b} (bid price) 4.95%
- r_t^{o} (ask price) 5.05%

- What is the price of the FRA, i.e., the FRA interest rate r_{FRA}?

Content

Futures and forwards

Hedging with futures and forwards is quite simple. Futures and forwards correspond to a simple purchase transaction where the traded product is paid for and delivered at a later date. The hedging is done by fixing the future price already at the time of trading. Futures and forwards are, therefore, also called forward transactions:

- Which include the obligation.
- The underlying or the base value.
- At a price agreed in advance (the future or forward price).
- At a date already fixed in advance.
- In a specified quality and quantity (contract) to take (long) or deliver (short)

The forward transaction is an obligation and must be fulfilled. This is why it is also referred to as an unconditional forward transaction. The period between the conclusion of a futures and a forward contract and the delivery date is called the term.

What is the difference between a future and a forward?

A *future* is a contract whose components are standardized:

- In terms of tradable sized units (contract size)
- With regard to the terms of payment and delivery

Futures are therefore tradable on the exchanges. Due to these characteristics, the future is transferable at any time and can also be accepted by a third party.

The counterpart to this is a *forward.* A forward allows an individual contract solution between two parties. This allows the trading partners a much greater degree of flexibility with regard to the contract size or delivery date. These types of individualized transactions are called over-the-counter or OTC transactions for short. The disadvantage of forwards is that they cannot be easily transferred to a third party. While the futures are traded daily on the futures exchanges and can thus be bought and sold by the contracting parties, the forwards are difficult to trade and remain as a contract between the two parties until they are fulfilled.

From a risk management perspective, there are also differences between the futures and the forwards. In the case of forwards, there is a settlement risk, as

(continued)

there is no exchange as an intermediary between the trading partners. Collateral is also not usually provided for forward transactions.

In the case of an interest rate futures transactions in the money market area, a distinction is made between

1. Money market futures that are exchange-traded.
2. Forwards that are traded OTC.

Money market futures:

In addition to the above-mentioned characteristics of futures, money market futures have the following special features:

1. Underlying: In the money market area, the underlying is not a government bond, but an agreed interest rate for a specific term. In the case of the EURIBOR future on the EUREX, for example, it is the EURIBOR interest rate for the 3-month forward.
2. Quotation: The future's price reflects the amount of an agreed interest rate. However, the quotation is made in such a way that the corresponding interest rate is subtracted from 100%. For example, if the interest rate is 1.5%, the future's price is 98.5% (=100% − 1.5%). As with bonds, the lower the interest rate, the higher the future's price. This means that the buyer of a money market future benefits from the falling interest rates, as this increases the price of the money market future. Similarly, the seller of a money market future benefits from the rising money market interest rates, as in this case the prices fall.
3. Fulfilment: As a physical delivery is not possible due to the underlying, the fulfilment in the case of interest rate futures transactions in the money market area always takes the form of a cash settlement.

Money market futures are often used to hedge an existing interest rate risk:

Money market futures from the buyer's perspective: From the buyer's perspective, money market futures represent an opportunity to make a *cash investment* at an agreed interest rate with an agreed maturity. The purchase of the money market future represents a strategy to hedge against an expected decline in the interest rates. The expected profit in the futures position resulting from the fall in interest rates exactly offsets the financial disadvantage that arises later when investing money due to lower interest rates. This creates an arrangement, in which the company is placed in such a position as if it would actually make a cash investment on the later date of the cash investment at the investment interest rate agreed today.

Money market futures from the seller's perspective: From the seller's perspective, on the other hand, the money market futures represent an opportunity to take out a *loan* at an agreed interest rate with an agreed term from the maturity date. The selling of the money market futures represented a strategy

(continued)

to hedge against an expected rise in the interest rates. The expected profit in the futures position resulting from the rise in the interest rates exactly offsets the financial disadvantage that arises later, when taking out a loan due to the higher interest rates. This creates an arrangement, in which the company is placed in such a position as if it would actually receive the loan on a later date of the borrowing at the loan rate agreed today.

In principle, the interest rate risk can be *hedged exactly in the case of money market* futures, since in the case of money market futures, the underlying represents an interest rate. However, an exact hedge requires that the contract sizes specified by the exchange, the term of the future and the maturities of the credit match the needs of the hedger.

Forward Rate Agreement (FRA):

The money market futures have the disadvantage that they are not flexible. They are only suitable for hedging if the contract sizes specified by the exchanges, the maturities of the futures and the maturities of the loans or money deposits match the hedger's needs.

This disadvantage can be resolved by individually agreed OTC transactions. These are known as forward rate agreements (FRAs). An FRA can be translated as "agreement on a future (interest) rate." In terms of the content, they are similar to money market futures, but, because of the free contractual agreement, some designations are different.

In a forward rate agreement, it is agreed to pay:

- A specific interest rate (FRA interest = forward interest).
- At a point in time, already determined today, from which an interest rate hedge is to begin (lead time period).
- For a certain future period (interest rate hedging period = FRA period).
- For a certain nominal amount (amount of money agreed).

Furthermore, the reference interest rate is determined, which is used to calculate the cash settlement (settlement payment). In the euro area, this is usually referred as EURO-LIBOR or EURIBOR.

In this context, the aforementioned contractual components (e.g., start and maturity date and interest rate or volume) can be agreed individually between the contracting parties.

FRA from the buyer's perspective: The buyer of an FRA secures a fixed interest rate for *borrowing for* the FRA period. The buyers of an FRA, thus, fear an increase in the interest rates.

FRA from the seller's perspective: The seller of an FRA secures an interest rate for the *investment of money*. The sellers of an FRA, thus, fear a decline in the interest rates.

Forward rate agreements use the following notation: A 3×9 FRA denotes an FRA that starts in 3 months (lead-in period) and has a total maturity of 9 months. The FRA period is, thus, the difference between the total term and the lead period, i.e., 6 months.

Important Formulas

Money Market Futures:

To determine the necessary number of money market futures for hedging purposes, the amount of the relevant cash position must be divided by the contract size of the money market future (contract value):

$$\boxed{\text{Hedge Ratio} = \text{Number of contracts} = \frac{\text{Nominal value spot position}}{\text{contract value}}} \tag{1}$$

The basis point value is required to calculate the cash settlement. This is calculated with the following formula:

$$\boxed{\text{Basis point value} = \text{Basis points} \cdot \frac{\text{Lead period in days}}{\text{FRA-Period in days}} \cdot \text{Notional amount}} \tag{2}$$

This can, then, be used to calculate the profit/loss (the net present value settlement) for the money market futures:

$$\boxed{\text{Profit/Loss} = \pm \text{Price change in Basis points} \cdot \text{Basis point value} \cdot \text{Number of contracts}} \tag{3}$$

Forward Rate Agreement:

The calculation of the gain/loss (cash value adjustment) in the case of a FRA is based on the interest rate difference related to the agreed FRA period and to the agreed cash amount, i.e.,

$$\boxed{\text{Profit/Loss} = \pm \text{Nominal Value} \cdot (\text{Reference Rate} - \text{FRA Interest Rate}) \cdot T_{\text{FRA}}} \tag{4}$$

T_{FRA} Duration of the FRA period

The settlement amount calculated below relates to the end of the term. The settlement payment is already paid out at the beginning of the FRA period and not at the end of the interest period. This is usual for interest payments. Hence, we have to discount this amount with the current interest rate.

$$\boxed{\text{Compensation payment} = \pm \frac{\text{Nominal value} \cdot (\text{Reference interest rate} - \text{FRA Interest rate}) \cdot T_{\text{FRA}}}{1 + \text{Reference interest rate} \cdot T_{\text{FRA}}}} \tag{5}$$

The pricing of the forward rate agreements is based on the following considerations. The purchase of an FRA to hedge an interest rate increase can be divided into two transactions:

1. We first take a loan over the entire term of the FRA. The term of the loan thus comprises the lead-up period plus the FRA period. This borrowing ensures that we can dispose of the desired loan amount in the FRA period at an interest rate already agreed today.
2. However, since we will not need the loan until later during the FRA period, we will invest the money that we do not need in the money market during the run-up period.

Thus, the purchase of an FRA is nothing more than a borrowing over the entire life of the FRA plus a cash investment made at the same time for the run-up period.

Therefore, it must always be possible to derive the interest rate for the total term from the interest rates for the lead period and the FRA period. The following formula applies:

$$\boxed{(1 + r_{t+T})^{t+T} = (1 + r_t)^t \cdot (1 + r_{(t,T)})^T} \tag{6}$$

r_{t+T} = Interest rate for the total term
$t + T$ = Total running time
r_t = Interest rate for the previous period
$_P t$ = Previous period
$r_{(t;T)}$ = FRA interest rate$_{(t, T)}$
$T_{FRA} = T$ = Transit time FRA

If we solve the equation for r_{FRA} we get the formula for the FRA interest rate:

$$\boxed{r_{(t,T)} = \sqrt[T]{\frac{(1 + r_{t+T})^{t+T}}{(1 + r_t)^t}} - 1} \tag{7}$$

Taking into account the bid and the offer rates, the formula for the FRA interest rate is as follows:

$$\boxed{r_{(t,T)} = \sqrt[T]{\frac{(1 + r^o_{t+T})^{t+T}}{(1 + r^b_t)^t}} - 1} \tag{8}$$

Workbook: `Case Study Risk Management_Part 2` Worksheet: ➲ `Hedging with Money Market Futures, FRA`

Money Market Futures:

The calculation of the *hedge ratio with a money market future in* the Excel is:

Excel example: `C11=C5/C6`

The calculation of the *base point value in* the Excel is:

Excel example: `C20='Assumptions money market future, FRA'!E9*C7/'Assumptions money market future, FRA'!C15*C5`

The calculation of the *profit/loss (cash value adjustment) for money market futures* in the Excel is:

Excel example: `C30=C5*C7/'Assumptions money market future, FRA'!C15*C28*C11`

Forward Rate Agreement:

The calculation of the *gain/loss (net present value adjustment) on an FRA* in the Excel is:

Excel example: `C55 =C43*(C50-C49)*C44/'Assumptions money market future, FRA'!C15`

The calculation of the *FRA interest rate in* the Excel is:

Excel-Example: `C90=((1+C87)^C83/(1+C88)^C84)^(1/C85)-1`

The calculation of the price of the FRA, i.e., the *FRA interest rate taking into account the bid price and the ask price* in the Excel reads:

Excel-Example: `C110=((1+C106)^C101/(1+C107)^C102)^(1/C103)-1`

Procedure

Hedging with the money market futures

- A company has a liquidity requirement of EUR 10 million in 6 months. The company, then, needs this liquidity for 3 months.
  - The relevant cash position is: EUR 10 million
  - The contract size for the 3-M EURIBOR Future is: EUR 1 million
- The company fears a rise in the interest rates and would like to hedge the interest rate risk with the money market futures.
- The first step is to calculate the hedge ratio. To do this, first transfer the nominal value of the spot position and the contract size for the 3-M EURIBOR future from the worksheet `Assumptions Money Market Future, FRA` to the worksheet `Hedging with Money Market Future, FRA`.
- To calculate the hedge ratio or the number of contracts, the amount of the relevant position is divided by the contract size of the money market future `C11=C5/C6`. In the case of the given company, 10 contracts of the EURIBOR future must, therefore, be sold at maturity in 6 months.
- Let us assume that the 3-M EURIBOR, on the day the future was purchased, was 4.0%. Let us now assume that after 6 months the interest rates of the 3-M-EURIBOR have risen by 200 bps to 6.0%.
- In the first step, the basis point value is calculated `C20='Assumptions money market future, FRA'!E9*C7/'Assumptions money market future, FRA'!C15*C5`. It results in 250.
- Based on this, the profit from the money market future can be calculated as follows `C21=C20*'Assumptions Money Market Future, FRA'!C11`. It amounts to EUR 50,000.
- The loss from the cash position corresponds to the gain from the money market future and is calculated as follows `C29=C5*C7/'Assumptions money market future, FRA'!C15*C28` The financial disadvantage (loss) compared to the 3-M-EURIBOR on the day of the future purchase is EUR 50,000.
- This allows the company to compensate for the financial disadvantage of borrowing. This is as a result of the increase in the borrowing rates by making an equally high profit in the futures position.
- If the company was wrong in its assessment and the 3-month interest rates were to fall to 3% against the expectations, the consequence would be that the loss from the money market future `C35=C20*'Assumptions money market future, FRA'!C17` would be offset by the financial advantage from the spot position against the 3-M EURIBOR on the day the future was purchased (`C36=C5*C7/'Assumptions money market future, FRA'!C15*C34`).

Hedging with the Forward Rate Agreement

- The initial scenario for hedging with the forward rate agreement (FRA) can be found in the problem definition. Since the company expects the interest rates to rise, it buys an FRA at the FRA interest rate of 4.0%. The 3-M-EURIBOR + 1% is set as the reference interest rate. Two days before the start of the FRA period, the 3-M-EURIBOR is 5.0%. Thus, the reference interest rate is fixed at 6.0%.

- With the given data, the profit or loss from the Forward Rate Agreement (FRA) can be calculated `C55=C43*(C50-C49)*C44/'Assumptions Money Market Future, FRA'!C15`. The profit from the FRA is EUR 25,000.
- The profit from the FRA is offset by a loss from the spot position for the amount of EUR 25,000 `C63=C43*C44/'Assumptions money market future, FRA'!C15*C62`.
- This allows the company to compensate for the financial disadvantage in borrowing as a result of the increase in the borrowing rates by making an equal profit in the FRA position.
- The settlement payment at the beginning of the FRA period is calculated by discounting with the current interest rate `C76=C63*C75`. The actual payout amount at the beginning of the forward is EUR 23,585.
- The price of the forward rate agreement (FRA) can be calculated using the total term, the previous period, the FRA period, and the total term and previous period interest rates `C90=((1+C87)^C83/(1+C88)^C84)^(1/C85)-1`. It is 6.5%, which is higher than the total term and the previous period's interest rates, as expected.
- The example can now be extended to include the bid price and the ask price. This changes the formula as follows: `C110=((1+C106)^C101/(1+C107)^C102)^(1/C103)-1`. This procedure increases the forward rate agreement (FRA) price to 6.8%.

Result (Figs. 2, 3, and 4)

	A	B	C
1			
2		**Money Market Futures**	
3			
4		A company has a capital requirement of EUR 10,000,000 (in EUR).	
5		The relevant spot position in EUR is:	10,000,000
6		The contract size for the 3-M-EURIBOR-Future is:	1,000,000
7		The company needs this liquidity for three months (= 90 days).	90
8			
9		**Calculation of the Hedge-Ratio**	
10			
11		Hedge ratio with reference to the number of contracts:	10
12			
13			
14		**Calculation of the Basis point value and the Profit from the Money Market future**	
15			
16		Let's assume that the 3-M-EURIBOR on the day of the futures purchase was:	4.0%
17		Now, let's assume that the interest rates of the 3M-EURIBOR have risen by the two basis points after six months:	2.0%
18		The 3-M-EURIBOR on the borrowing date is:	6.0%
19			
20		Base Point value:	250
21		Profit from the Money Market Future:	50,000
22		Profit from the money market future, taking into account the number of contracts	500,000
23			
24		**Calculation of the profit/loss from the Spot position**	
25			
26		Spot position after the expiry of the futures, at which the company would like to take the desired 3M-EURIBOR.	
27			
28		Interest rate increase from the starting point to the date of borrowing:	2.0%
29		Financial disadvantage (loss) compared to the 3-M-EURIBOR on the day of the futures purchase.	50,000
30		Loss taking into account the number of contracts	500,000
31			
32		**Situation when three-month interest rates fall to 3%**	
33			
34		Interest rate decline from the starting point to the date of borrowing:	1.0%
35		Loss from the Money Market Future:	25,000
36		Financial advantage (profit) compared to the 3M-EURIBOR on the day of the futures purchase.	25,000
37			

Fig. 2 Money market futures

	A	B	C	D
40				
41		**Forward Rate Agreement (FRA)**		
42				
43		A company has a liquidity requirement of EUR 10 million in six months.	10,000,000	Euro
44		The company then needs this liquidity for three months.	90	days
45				
46		Based on its rating, the company would currently have to pay the 3M-EURIBOR plus a Risk Premium of one percent for a three-month loan.		
47				
48		Since the company expects interest rates to rise, it buys an FRA at the FRA interest rate of 4.0%.	4.0%	
49		The 3M-EURIBOR + 1% is set as the reference interest rate	5.0%	
50		Two days before the start of the FRA period, the 3M-EURIBOR is 5.0%. This fixes the reference interest rate at 6.0%.	6.0%	
51				
52				
53		**Profit/loss from the FRA**		
54				
55		Profit from the FRA:	25,000	
56				
57				
58		**Profit/loss from the Spot position on the Borrowing**		
59				
60		Spot position after the expiry of the futures, to which the company would like to take the desired 3M EURIBOR.		
61				
62		Interest rate increase from the starting point to the date of borrowing:	1.0%	
63		Financial disadvantage (loss) versus the 3M-EURIBOR on the day of the FRA purchase.	25,000	
64				
65				
66		**Actual payout amount at the beginning of the Forward**		
67				
68		In our example, the maturity of the FRA is 30 days.		
69				
70		The compensation amount calculated above refers to the end of the term.		
71				
72		Since the compensation payment is already paid out at the beginning of the FRA period and not at the end of the interest period, as is usual with the interest payments, we have to discount this amount using the current interest rate.		
73				
74		Discount rate:	6.0%	
75		Discounting factor:	0.9434	
76		Actual payout amount at the beginning of the forward:	23,585	

Fig. 3 Forward Rate Agreement—Part 1

Literature and References to the Excel Tool

Ernst, D., Häcker, J. (2021): Risk management in business step by step, UTB, Course 3—Assignment 1.

Bösch, M. (2020): Derivate: Verstehen, anwenden und bewerten, 4th edition, Munich, pp. 215-224.

Hull, J. C. (2014): Risk management: banking, insurance and other financial institutions, 3rd edition, Pearson, pp. 283-295.

See Excel file `Case Study Risk Management Part 2`, Excel worksheet `Hedging with money market futures, FRA`

	A	B	C	D
78				
79		**Calculation of the price of the FRA i.e. the FRA interest rate**		
80				
81		The following initial values are given:		
82				
83		t+T the total running time:	0.75	
84		t the previous period:	0.50	
85		T the term of the FRA period:	0.25	
86				
87		$r_{(t+T)}$	5.5%	
88		r_t	5.0%	
89				
90		Price FRA = $r_{(t;T)}$	6.5%	
91				
92				
93		**Calculation of the price of the FRA, i.e. the FRA interest rate taking into account the Bid price and the Ask price**		
94				
95		In our example, we disregard the fact that the interest rates on an investment differ from the interest rates on a loan.		
96		The terms bid price and ask price are also used for this purpose.		
97		The same also applies to an FRA, where the FRA interest rate in the purchase will be higher than the interest rate in the sale.		
98				
99		The following initial values are given:		
100				
101		t+T the total running time:	0.75	
102		t the previous period:	0.50	
103		T the term of the FRA period:	0.25	
104				
105		r_{t+T} the total running time:	5.45%	
106		r_{t+T}^{o} the total running time:	5.55%	
107		r_{tb} the previous period:	4.95%	
108		r_t the previous period:	5.05%	
109				
110		Price FRA = $r_{(t;T)}$	6.8%	
111				

Fig. 4 Forward Rate Agreement—Part 2

Assignment 19: Hedging Interest Rate Risk Using Interest Rate Swaps

Task

Calculation of the swap point

- A company wants to hedge an amount of EUR 1,000,000 against an interest rate risk with an interest rate swap.
- The spot interest rates are:

  t(0) = 1.6%; t(1) = 2.0%; t(3) = 2.5%; t(4) = 2.8% and t(5) = 3.1%.

- Calculate the fair swap rate.

Modelling of the interest rate swap via the sale of a fixed-rate bond combined with the purchase of a floater

- The variable interest rates are:

  t(0) = 2.1%; t(1) = 3.0%; t(3) = 2.0%; t(4) = 2.2% and t(5) = 2.5%.

- Model the interest rate swap from the payer perspective via the sale of a fixed-rate bond combined with the purchase of a floater.
- What is the value of the interest rate swap?

  Modelling of the interest rate swap as a sum of the forward transactions

- Calculate the FRA interest rates *r(t;1)* for the years t(1)–t(5).
- Model the interest rate swap as a sum of the forward transactions.
- What is the value of the interest rate swap?

Content

A swap is an unconditional forward contract typically traded over-the-counter. A swap is an agreement between

- Two parties
- For exchanging payment flows
- In the future

When concluding the contract, three things need to be established:

- The Term of the swap.
- The method of calculating the cash flows.
- The dates on which these cash flows are to be exchanged between the contracting parties.

Advantages of the swaps include:

- The procurement of favorable financial resources.
- The reducing costs of export credits.
- The financing of foreign branches.
- Facilitating the raising of capital.

In the context of risk management, the swap holders want to use the swap to exchange an undesirable risk position for a desirable one.

In purely formal terms, a swap is a bilateral financial contract that represents a payment stream between two parties. An exchange of cash flows is thus agreed, which is preceded by a nominal amount and the conditions fixed at the time of conclusion.

Due to their design, the *interest rate swaps* are unconditional forward transactions. They are regarded as established instruments of the futures market and are used as part of an operational interest rate risk management. Although, the interest rate swaps are over-the-counter instruments, the

(continued)

volumes traded are so high that the market is considered to be quite liquid. The banks are often considered as the market partners. They act either as arrangers or intermediaries.

An interest rate swap is when two parties agree to exchange the interest payment obligations in one currency for a fixed term. The interest payment obligations are based on an equal amount of capital. Since, the principal amounts involved are the same, they are not swapped. Only the way in which the interest is calculated differs. For example, a fixed interest rate in euros is swapped for a variable interest rate in euros.

- The fixed interest receiver refers to the swap accordingly, as a fixed interest receiver swap or receiver swap.
- The fixed rate payer refers to a fixed rate payer- or payer swap.

An interest rate swap is valued by replicating the interest rate swap using other financial instruments. The first way to value an interest rate swap is with bonds.

- *Valuation with bonds*: In simplified terms, one could say that an interest rate swap is analogous to the exchange of a bond with a fixed interest rate for a bond with a variable interest rate (floater). In an interest rate swap, only the interest rate flows are "geneticized." The payer either assumes that interest rates will rise faster than the market expectations or that they will fall more slowly than expected. Based on this assumption, he pays the fixed rate. His antagonist is the receiver. Should the payer's market expectation come to a pass, he makes a profit; conversely, he realizes a loss and the receiver a profit. Thus, the swap (like other derivatives) is a classic zero-sum game.
- *Valuation by forwards*: In simplified terms, however, one could also say that in a swap, the future payments are exchanged on predefined dates at the payment terms that have also been agreed. This procedure corresponds to a forward. A swap can, therefore, also be interpreted as a sum of forward transactions.

Important Formulas

The formula for determining the fair swap rate is as follows:

$$\boxed{s_T = \frac{1 - \frac{1}{[1+r(0,T)]^T}}{\frac{1}{[1+r(0,1)]^1} + \frac{1}{[1+r(0,2)]^2} + .. + \frac{1}{[1+r(0,T)]^T}}} \tag{9}$$

s_T Fair swap rate
$r(0, T)$ FRA interest rate at time t for term T

We have already learnt the formula for determining the FRA interest rate. It is:

$$r_{FRA} = r_{(t;T)} = \sqrt[T]{\frac{(1 + r_{t+T})^{T_{t+T}}}{(1 + r_t)^{T_t}}} - 1 \tag{10}$$

Workbook: `Case Study Risk Management_Part 2` Worksheet: ➲ `Hedging with Interest Rate Swaps`

The calculation of the fair swap rate in the Excel is:

Excel-Example: `C9=(1-1/(1+H8)^H7)/(1/(1+D8)^D7+1/(1+E8)^E7+1/(1+F8)^F7+1/(1+G8)^G7+1/(1+H8)^H7)`

The calculation of the FRA interest rate in the Excel is:

Excel-Example: `E49=((1+E48)^E47/(1+D48)^D47)^(1/(E47-D47))-1`

Procedure

Payer interest rate swap: sale fixed-rate bond plus purchase floater

- We know that from the payer's perspective, the value of an interest rate swap can be replicated by selling a fixed-rate bond and simultaneously buying a floater. The payer sells a fixed-rate bond to the receiver. He receives a fixed interest rate and pays the receiver a floating interest rate in return.
- The coupon of the fixed-rate bond is equal to the agreed swap rate s_T. Furthermore, we are aware that the fair value of an interest rate swap is zero when the contract is concluded.
- The swap amount is EUR 1,000,000 (cell `C4`).
- Next, we take the spot `interest` rates for years t(1) to t(5) from the `interest rate swap assumptions`, which are subsequently used to calculate the discount factors.
- In cell `C9`, we calculate the swap rate at which the swap is "fairly" valued at contract inception:

  `C9=(1-1/(1+H8)^H7)/(1/(1+D8)^D7+1/(1+E8)^E7+1/(1+F8)^F7+1/(1+G8)^G7+1/(1+H8)^H7).`

  In our example, the swap rate is 3.0590%. This means that the swap does not benefit either side and takes on a value of zero.
- By selling the fixed-rate bond, the company has a cash inflow of EUR 1,000,000 in t(0) (cell `C10`), which must be repaid in year t(5) (cell `H10`).

- The interest payable by the entity is based on the swap rate and the swap amount `D11=-$C$9*$C$10`.
- With the discounting factor `D13=1/(1+D8)^D7`, the payments of the fixed interest side are discounted `D14=D12*D13`.
- The sum of the present values of the disbursements (cell `C14`) corresponds exactly to the value of the deposit (cell `C10`).

- Now let us look at the receiver view. The receiver buys a floating rate bond (floater) and exchanges a floating rate for a fixed rate.
- The best way to imagine this is that a floater of EUR 1,000,000 is purchased each year (cells `C19:G19`) and repaid the year after, including the interest (cells `D22:H22`).
- Cells `D18:H18` show the interest rates for the variable interest payments of the floater. When the contract is concluded, only the first interest rate of 1.6% is known.
- The purchase of the floater in the year 0, therefore, results in a payment of EUR 1,016,000 (cell `D22`). Since the variable interest rate is the same as the discount rate, the present value is EUR 1,000,000 (cell `D23`).
- At the end of year t(1), the floating interest rate for year t(2) is fixed. It amounts to 3.00% (cell `E18`). This results in a deposit of EUR 1,030,000 in the form of the interest and the repayment. Since the discount rate is also 3.00%, this again results in a present value of 1,000,000. We see here that it does not matter what the variable interest rate is. The present value of the floater is always 1,000,000 on the interest dates. We proceed in the same way in years t(3) to t(5).
- If we add the sum of the present values of the fixed-rate bond (EUR −1,000,000) and the value of the floater (EUR 1,000,000), we get a value of zero. When the contract is concluded, the values equal each other and the interest rate swap has a value of zero.

- As a control calculation, we can alternatively determine the same result by adding the present values of the payments of the fixed-rate side and the floater (cells `C27:H27`). At contract inception, i.e., in t(0), the two payments cancel out. For years t(1)–t(4), only the values of the fixed-rate side are relevant, since the present value of the floater is always equal to the payment for the floater for the new year. In the fifth year, no more floaters are purchased. Therefore, the NPV of the fixed-rate side of EUR −884,693 compares with the NPV of the floater of EUR 1,000,000. The difference in the net present values is EUR 115,307. If the values determined are added over the 5 years, the sum of the present values again results in a value of zero.

Payer interest rate swap: sale of the fixed-rate bond plus purchase of the forward rate agreement

- The valuation approach presented so far interprets an interest rate swap as the difference between the value of a floater and a fixed-rate bond. However, we can also interpret a swap as follows: In a swap, the future payments are exchanged on the specified dates. The calculation and payment parameters are determined in advance. This makes the swap similar to a forward transaction. We can, therefore, interpret a swap as a sum of forward transactions. In this approach, the swap valuation is carried out in the following steps:
- The variable payments are replaced by the corresponding FRA interest rates in cells (`D49:H49`). The FRA interest rate in t(1) in the cell `D49` corresponds to the spot interest rate in the cell `D48`, as this interest rate is not affected by the forward transaction. The FRA rates in t(2) through t(5) are calculated using the FRA valuation formula:

  `E49=((1+E48)^E47/(1+D48)^D47)^(1/(E47-D47))-1.`
- The discount factors in the cells (`D54:H54`) are calculated using the FRA rates.
- The capital value can now be calculated and is EUR 1,000,000.
- The capital value of the fixed-interest bond has already been calculated and amounts to EUR −1,000,000.
- The sum of the two principal values represents the value of the swap. Both procedures lead to the same result. The interest rate swap has a value of zero.

Payer interest rate swap: sale of fixed-rate bond plus purchase of a floater in the event of a change in the yield curve

- As we have seen, the swap has a value of zero at a fair swap rate.
- How does the situation change if the spot interest rates change after the contract has been concluded? Which of the two parties benefits, and which loses?
- Since the swap rate was agreed as fixed, the swap receiver has to pay it to the payer. In exchange, he receives a variable interest rate, which has increased due to the rise in the interest rates.
- For the payer, the increase in the interest rates does not change anything, since the value of the floater is always the same as the nominal value at the interest rate dates.
- However, changes in the spot rates affect the receiver and the value of the fixed-rate bond because the swap rate is fixed and the change in spot rates creates a mismatch with the initial situation.
- If the spot interest rates rise, the capital value of the fixed-rate bond falls, and vice versa.

- From the receiver's point of view, there is a loss because he receives the lower swap rate from the payer while the market rate has increased.
- From the payer's point of view, on the other hand, the value of the swap increases when interest rates rise, because the payer receives the loan from the receiver at more favorable conditions than are available on the market.
- In our example, we simply replace the previous spot interest rates with the spot interest rates that have increased by 0.5% in the cells `E72:H72` when the receiver sells the fixed-rate bond.
- However, the interest payment to the receiver (cells `E75:H75`) is based on the agreed swap rate (cell `D73`), which was calculated on the basis of the old spot interest rates.
- These interest payments are discounted using the discount factors (cells `E77:H77`) based on the new spot interest rates.
- When compared to the initial situation, this results in lower present values (cells `E78:H78`) and also a lower net present value (cell `D78`). The net present value of the fixed-interest side is EUR −992,146.
- Since the net present value of the floater has remained the same at EUR 1,000,000, from the receiver's perspective, there is a loss of EUR 7,854 (cell `D89`). From the payer's point of view, the development is positive. He makes a profit of EUR 7,854.

Result (Figs. 5, 6, and 7)

Closing at Swap rate						
Swap amount	10,00,000					
Sale of the fixed-rate bond						
Years	0	1	2	3	4	5
Spot interest rates r(0;t)		1.60%	2.00%	2.50%	2.80%	3.10%
Swap rate	3.0590%					
Recording and repayment	10,00,000					-10,00,000
Interest charges		-30,590	-30,590	-30,590	-30,590	-30,590
Payments of the fixed interest side	10,00,000	-30,590	-30,590	-30,590	-30,590	-10,30,590
Discount factor		0.9843	0.9612	0.9286	0.8954	0.8584
Present value fixed interest side	-10,00,000	-30,108	-29,402	-28,406	-27,391	-8,84,693
Purchase of the floater						
Years	0	1	2	3	4	5
Variable Annual Interest Rate		1.60%	3.00%	2.00%	2.20%	2.50%
Buy floaters	-10,00,000	-10,00,000	-10,00,000	-10,00,000	-10,00,000	
Sale floaters		10,00,000	10,00,000	10,00,000	10,00,000	10,00,000
Interest charges		16,000	30,000	20,000	22,000	25,000
Floater payments		10,16,000	10,30,000	10,20,000	10,22,000	10,25,000
Present value floaters	10,00,000	10,00,000	10,00,000	10,00,000	10,00,000	10,00,000
Value of the Interest Rate Swap	0					
Present value Fix plus Floater	0	-30,108	-29,402	-28,406	-27,391	1,15,307
Check	OK					

Fig. 5 Payer interest rate swap: sale of fixed-rate bond plus purchase of floater

Closing at Swap Rate						
Swap amount	10,00,000					
Sale of the Fixed-rate Bond						
Years	0	1	2	3	4	5
Spot interest rates r(0;T)		1.60%	2.00%	2.50%	2.80%	3.10%
Swap rate	3.0590%					
Recording and repayment	10,00,000					-10,00,000
Interest charges		-30,590	-30,590	-30,590	-30,590	-30,590
Payments of the fixed interest side	10,00,000	-30,590	-30,590	-30,590	-30,590	-10,30,590
Discount factor		0.9843	0.9612	0.9286	0.8954	0.8584
Present value fixed interest side	-10,00,000	-30,108	-29,402	-28,406	-27,391	-8,84,693
Modeling via Forward Rate Agreement						
Years	0	1	2	3	4	5
Spot Interest Rates r(0;T)		1.60%	2.00%	2.50%	2.80%	3.10%
FRA Interest Rates r(t;1)		1.60%	2.40%	3.51%	3.71%	4.31%
Buy FRA(t;1)	-10,00,000	-10,00,000	-10,00,000	-10,00,000	-10,00,000	
Sale FRA(t;1)		10,00,000	10,00,000	10,00,000	10,00,000	10,00,000
Interest charges		16,000	24,016	35,074	37,053	43,088
Payments from the FRA(t;1)		10,16,000	10,24,016	10,35,074	10,37,053	10,43,088
Discount factor		0.9843	0.9765	0.9661	0.9643	0.9587
Present Value Forward Rate Agreen	10,00,000	10,00,000	10,00,000	10,00,000	10,00,000	10,00,000
Value of the Interest Rate Swap	0					
Present value FRA plus Floaters	0	-30,108	-29,402	-28,406	-27,391	1,15,307
Check	OK					

Fig. 6 Payer interest rate swap: sale of fixed-rate bond plus purchase of forward rate agreement

1 year after closing, the Yield Curve shifts by 50 basis points					
Swap amount	10,00,000				
Sale of the fixed-rate bond					
Years	0	1	2	3	4
Spot interest rates r(0;t)		2.10%	2.50%	3.00%	3.30%
Swap rate	3.0590%				
Recording and repayment	10,00,000				-10,00,000
Interest charges		-30,590	-30,590	-30,590	-30,590
Payments of the fixed interest side	10,00,000	-30,590	-30,590	-30,590	-10,30,590
Discount factor		0.9794	0.9518	0.9151	0.8782
Present value fixed interest side	-9,92,146	-29,961	-29,116	-27,994	-9,05,075
Purchase of the floater					
Years	0	1	2	3	4
Variable Annual Interest Rate		2.10%	2.50%	3.00%	3.30%
Buy floaters	-10,00,000	-10,00,000	-10,00,000	-10,00,000	
Sale floaters		10,00,000	10,00,000	10,00,000	10,00,000
Interest charges		21,000	25,000	30,000	33,000
Floater payments		10,21,000	10,25,000	10,30,000	10,33,000
Present value floaters	10,00,000	10,00,000	10,00,000	10,00,000	10,00,000
Value of the Interest Rate Swap	7,854				
Present value Fix plus Floater	7,854	-29,961	-29,116	-27,994	94,925
Check	OK				

Fig. 7 Payer interest rate swap: sale of fixed-rate bond plus purchase of floater in the event of a change in the yield curve

Literature and References to the Excel Tool

Ernst, D., Häcker, J. (2021): Risk management in business step by step, UTB, Course 3—Assignment 2.

Bösch, M. (2020): Derivate: Verstehen, anwenden und bewerten, 4th edition, Munich, pp. 251-268.

Hull, J. C. (2014): Risk management: banking, insurance and other financial institutions, 3rd edition, Pearson, pp. 283-295.

See Excel file `Case Study Risk Management Part 2`, Excel worksheet `Hedging with interest rate swaps`

Assignment 20: Hedging Interest Rate Risk Using Options (Caplet)

Task

We assume the same initial situation and the same initial data as in Assignment 19. However, the company now wants to design the hedge with the options (caplet).

Initial situation

- A company has taken out a loan in the amount of EUR 1,000,000 with a term of 5 years. The loan has a variable interest rate based on the 12-month EURIBOR. In order to hedge against the interest rate increases occurring in the meantime, the company wants to conclude a cap agreement with a bank with a strike price (interest rate cap) of 2.6% (= base rate). This means that the company would like to hedge against all the interest rate risks greater than 2.6%.
- The spot interest rates of the 12-month EURIBOR (= reference interest rate) are:
  - t(0) = 1.6%; t(1) = 2.0%; t(3) = 2.5%; t(4) = 2.8% and t(5) = 3.1%.

Calculation of the zero bond discount factors, the zero bond interest rates, and the FRA rates

- Calculate the zero bond discount factors.
- Calculate the zero bond interest rates.
- Calculate the FRA rates.

Compensation payments of the cap

- Calculate the compensation payments of the cap based on the reference interest rates and the base rate.

Intrinsic value of the cap

- Calculate the intrinsic value of the cap.

Cap price

- Calculate the price of the cap.

Content

Interest rate options are agreements that hedge interest rate risk by fixing an interest rate cap or floor. They are, therefore, also referred to as the interest rate caps/floors. Such hedging strategies are used for variable-rate loans or investments, whose interest rates can fluctuate over their term. This gives rise to interest rate risk. A distinction is made between caps, floors, and collars.

Cap: The buyer of an interest rate option hedges against rising interest rates by paying the option premium to the seller. From a borrower's point of view, the upward limitation of interest rates (cap) is economically relevant in the case of variable interest rates. The option seller, also known as the writer, pays the option buyer a compensation payment equal to the difference between the reference interest rate (e.g., Libor or EURIBOR) and the underlying rate for the underlying loan amount if the base rate (cap) is exceeded.

Floor: The buyer of an interest rate option hedges against the falling interest rates by paying the option premium to the seller. An investor is interested in a downward interest rate threshold (floor) on a floating-rate investment. If the interest rate falls below the base rate (upper/lower limit), the seller of the option pays the buyer of the option a compensation payment, i.e., equal to the difference between the reference interest rate (e.g., Libor or EURIBOR) and the underlying rate (exercise price) for the underlying investment amount.

Collar: If the buyer combines both the forms of contract (interest rate cap and a floor), an interest rate spread called a collar is created. An interest rate collar is generated by buying the interest rate caps and selling the interest rate floors. The following applies:

- The maturity and the nominal and underlying values of the cap and the floor are equal.
- The exercise price of the cap is higher than that of the floor.
- The purchase of the cap protects the company against the rising interest rates.
- The sale of the floor does not allow the company to benefit from the interest rate reductions above the agreed limit, but generates a premium income that reduces or offsets the premium payable.

In this assignment, we will deal with an interest rate cap. An interest rate cap is not a single interest rate option, but a series of consecutive European interest rate options. The options for the individual interest rate periods are called the *caplets*.

(continued)

When calculating the caps, we use the models with continuous variables, since a continuous model is used in particular when calculating the price. Furthermore, the models require us to work with forward rate agreement (FRA) rates based on the zero coupon rates (zero bond rates) and the zero bond discount factors (zero coupon discount factors).

Zero coupon rates are the interest rates that integrate the compound interest effect for the multi-period investments and exclude the payment of interim interest. The zero coupon interest rates for different points in time form the *yield curve*. The zero-coupon interest rates can be used to derive the *zero-coupon discount factors* and vice versa. The zero-coupon discounting factors can, in turn, be used to calculate the *forward rate agreement (FRA) interest rates.*

The *compensation payment* for the respective interest period of the caplet corresponds to the difference between the reference interest rate on the fixing date (at the beginning of the interest period) and the base interest rate, based on the contract volume and the length of the interest period.

As with all the other options, the *intrinsic value of* the cap can also be calculated without an option pricing model. For this purpose, the reference interest rates, which are still unknown from today's perspective, are replaced by the forward rate agreement (FRA) interest rates determined from the current yield curve.

A cap *is* valued using the *Black model* for the caps (please do not confuse this with the Black-Scholes model). The Black model was developed by Fischer Black in 1976 for options on futures. To use the Black model, the cap is first broken down into its caplets. The value of each caplet is then determined analytically by the Black model. The total value of the cap is the sum of the values of the individual caplets.

Important Formulas

The formula of the zero bond discount factors in t(0) (i.e., with the lead period $T_{VP} = 0$) is:

$$\boxed{\mathrm{ZB} - \mathrm{DF}_{0,T_{\mathrm{FRA}}} = \frac{1 - r_{0,T} \cdot \sum_{j=1}^{t+T-1} \mathrm{ZB} - \mathrm{DF}_{0,j}}{1 + r_T}} \tag{11}$$

$\mathrm{ZB} - \mathrm{DF} =$ Zero bond discount factor
$t + T =$ Total running time
$t =$ Previous period
$T_{\mathrm{FRA}} = T =$ Term FRA

You can then, use these zero bond discounting factors to calculate the zero bond discounting factors for a lead time period >0 based on the zero bond discounting factors with the valuation date $t = 0$ using the following formula:

$$\boxed{\mathrm{ZB}-\mathrm{DF}_{t,T} = \frac{\mathrm{ZB}-\mathrm{DF}_{(0,t+T)}}{\mathrm{ZB}-\mathrm{DF}_{(0,T)}}} \tag{12}$$

Now, the zero bond interest rates can be calculated based on the zero bond discount factors:

$$\boxed{r_{\mathrm{ZB}} = \frac{e^{\mathrm{ZB}-\mathrm{DF}_{(0,t+T)}}}{-(t+T)}} \tag{13}$$

Furthermore, the FRA interest rates can also be calculated based on the zero bond discount factors:

$$\boxed{r_{\mathrm{FRA}} = r_{(t,T)} = \frac{1-\mathrm{ZB}-\mathrm{DF}_{(t,t+T)}}{\sum_{n=1}^{t+T}\mathrm{ZB}-\mathrm{DF}_{(t,n)}}}$$

The formula for the compensation payment for the respective interest period of the caplet is as follows:

$$\boxed{\begin{aligned}&\text{Compensation Payment}\\&= \text{Nominal Value}\cdot(\text{Reference Interest Rate}-\text{Base Interest Rate})\\&\cdot\text{Hedging Period} = NV\cdot(RI-X)\cdot HP\end{aligned}} \tag{14}$$

NV = Nominal value, contract volume
RI = Reference interest rate
X = Base interest rate
HP = Hedging period= term

The intrinsic value of the caplets is calculated using the following formula:

$$\boxed{\begin{aligned}&\text{Intrinsic Value of a Caplet} =\\&= \text{Nominal Value}\cdot\text{Hedging Period}\cdot(\text{Forward Rate}-\text{Base Interest Rate})\\&\cdot e^{-r\cdot\text{Total Running Time}}\\&= NV\cdot HP\cdot(\mathrm{FR}-X)\cdot e^{-r\cdot TRT} = NV\cdot HP\cdot(\mathrm{FR}-X)\cdot\mathrm{ZB}-\mathrm{DF}\end{aligned}} \tag{15}$$

FR = Forward rate
X = Base interest rate
r = Zero bond interest rate or zero-coupon interest rate

$t+T$ = Total running time
ZB − DF = Zero bond discount factor or zero-coupon discount factor

The formula of the Black model is:

$$\boxed{\text{Caplet} = NV \cdot HP \cdot e^{-r \cdot t+T} \cdot [\text{FR} \cdot N(d_1) - X \cdot N(d_2)]} \tag{16}$$

Thereby are d_1 and d_2 are defined as:

$$\boxed{d_1 = \frac{\ln(\frac{\text{FR}}{X}) + \sigma^2 \cdot \frac{T}{2}}{\sigma \cdot \sqrt{T}}} \tag{17}$$

$$\boxed{d_2 = \frac{\ln(\frac{\text{FR}}{X}) - \sigma^2 \cdot \frac{T}{2}}{\sigma \cdot \sqrt{T}} = 1 - \sigma \cdot \sqrt{T}} \tag{18}$$

$N(d)$ = Distribution function of the standard normal distribution
T = Time left til maturity

Workbook: `Case Study Risk Management_Part 2` Workshe➲`Hedging with Caps`

The calculation of the *zero bond discount factor* in t(0) in the Excel is:

Excel-Beispiel: `F14=(1-F32*(D14+E14))/(1+F32)`

The calculation of the *zero bond discount factor* for a lead period >0 and a term of 3 years based on the zero bond discount factors in the Excel is as follows:

Excel example: `F15=G14/D14`

The calculation of the *zero bond interest rates* based on the zero bond discount factors in the Excel reads:

Excel-Beispiel: `F25=LN(F14)/-F13`

The calculation of *FRA interest rates* based on the zero bond discount factors in the Excel reads:

Excel-Beispiel: `F33=(1-F15)/(D15+E15+F15)`

The calculation of the *compensation payment* in the Excel reads:

Excel example: F43=E43*'Assumptions Hedging with Caps'!$D $14

The calculation of the *intrinsic value* in t(0) in the Excel is:

Excel example: G55=F55*G14

The calculation of the *prices of the caplets* in the Excel is:

Excel example: N65=If(C65=0;065;IF(D65<>-1;'Assumptions Hedging with Caps'!D14*G14*(E65*I65-F65*J65);0))

Procedure

Calculation of zero bond discount factors

- In the first step, the spot interest rates are taken from the assumptions hedging with the caps (cells D7:H7).
- The zero bond discount factors of the year t(0) are then calculated (cells D14:H14). Please note that each of the cells contains a different formula type.
- Using the calculated zero bond discount factors, the zero bond discount factors for the next 4 years with their different maturities can now be calculated (cells D15:H18).

Calculation of the zero bond interest rates

- Based on the zero bond discount factors of the year t(0), the zero bond interest rates can now be calculated (cells D25:H25). For the term of 3 years, the Excel formula is F25=LN(F14)/-F13.

Calculation of the FRA interest rates

- Based on the zero bond discount factors of year t(0), the FRA interest rates can now also be calculated (cells D32:H36). Please note that each of the cells contains a different formula type.

Calculation of the compensation

- To calculate the compensation payment, the base interest rate is first taken from the assumptions hedging with caps (cells C43:D66). The reference interest rate is the spot interest rate in year t(0) with maturities of 2–5 years (cells D53:E56). If the reference interest rate exceeds the base interest rate, the option is exercised E45=IF(D45>C45;D45-C45;0). The settlement payment is determined by the calculated interest rate difference and the contract volume F45=E45*'Assumptions Hedging with Caps'!$D$14.

Calculation of the intrinsic value of the caplets

- To calculate the intrinsic value of the caplets, the FRA interest rates FRA(t,1) already calculated are taken over (cells C53:C56), as is the base interest rate (cells D53:D56). Based on this, the interest rate difference between the FRA interest rates and the base interest rate is calculated E55=IF(C55<D55;0; C55-D55). The intrinsic value of the caplet at the end of the total term is calculated by multiplying the determined interest rate difference by the contract volume F55=E55*'Assumptions Hedging with Caps'!$D$14. The intrinsic value of the caplet in t(0) is calculated by multiplying the already determined intrinsic value of the caplet at the end of the total term by the zero bond discount factor G55=F55*G14.

Calculation of the option price

- First, the caplet number, the lead time, and the total term are taken from the assumptions Hedging with Caps (cells B63:D66). Furthermore, the already calculated FRA interest rates FRA(t,1) (cells E63:E66) and the base interest rate (cells F63:F66) are transferred. Again, the interest rate difference between the FRA interest rates and the base rate is calculated G65=IF (E65<F65;0;E65-F65). Furthermore, the volatility (cells K63:K66) is taken. In the cells I63:J66, *N(d1)* and *N(d2)* are calculated. *N(d1)* is the delta of the option and *N(d2)* is the probability that the option will be exercised; that is, *N(d2)* is the probability that the *FRA* will be greater than *X. The* price of each caplet is calculated using the option pricing formula above N65=IF (C65=0;O65;If(D65<>-1;'Assumptions Hedging with Caps'! D14*G14*(E65*I65-F65*J65);0)). The total price of the cap is equal to the sum of the prices of the caplets N68=SUM(N63:N66). It is EUR 37,674. In (cells O63:P66), the prices of the caplets are divided into their intrinsic value and their time value.

Result (Figs. 8 and 9)

Determination FRA Interest Rates using Zero-Bond Discount factors

Spot Interest Rates r(0;T)	1	2	3	4	5
t(0)	1.60%	2.00%	2.50%	2.80%	3.10%

Zerobond Discount Factors

		Term in Years				
		1	2	3	4	5
Maturity in the Ye	t(0)	0.98425	0.96109	0.92816	0.89450	0.85664
	t(1)	0.97647	0.94301	0.90881	0.87034	
	t(2)	0.96574	0.93071	0.89131		
	t(3)	0.96373	0.92294			
	t(4)	0.95767				

Zerobond Interest Rates

	Term in Years				
	1	2	3	4	5
t(0)	1.59%	1.98%	2.48%	2.79%	3.09%

FRA-Rate

		Term in Years				
		1	2	3	4	5
Maturity in the Ye	t(0)	1.60%	2.00%	2.50%	2.80%	3.10%
	t(1)	2.41%	2.97%	3.22%	3.51%	
	t(2)	3.55%	3.65%	3.90%		
	t(3)	3.76%	4.08%			
	t(4)	4.42%				

Fig. 8 Determination of FRA interest rates via zero bond discount factors

Calculation of the Compensation Payment

Time	Base Rate	Reference Interest Rate	Difference between Interest Rates	Compensation Payment in EUR
t(1)	2.60%	2.00%	0.00%	0
t(2)	2.60%	2.50%	0.00%	0
t(3)	2.60%	2.80%	0.20%	2,000
t(4)	2.60%	3.10%	0.50%	5,000

Calculation of Intrinsic Value

Time	Forward Rate FR(t,1)	Base Rate (X)	Interest rate difference max (FR-X,0)	Intrinsic value of caplets in tGZ in EUR	Intrinsic value the caplets in t0 in EUR
t(1)	2.41%	2.60%	0.00%	0	0
t(2)	3.55%	2.60%	0.95%	9,480	8,799
t(3)	3.76%	2.60%	1.16%	11,637	10,410
t(4)	4.42%	2.60%	1.82%	18,196	15,587

Calculation of the Option Price

Caplet Nr.	Lead Time (VZ)	Total running time (GL)	Forward Rate FRA(t,1)	Base Rate (X)	Interest rate difference max (FR-X,0)		N(d1)	N(d2)	Volatility	d1	d2	Prices of the Caplets	Intrinsic Value	Time Value
0	1	2	2.41%	2.60%	0.00%		0.39	0.32	20.00%	-0.28017	-0.48	1,139.22		1,139
1	2	3	3.55%	2.60%	0.95%		0.89	0.83	20.00%	1.24048	0.96	9,343.14	8,799	545
2	3	4	3.76%	2.60%	1.16%		0.89	0.81	20.00%	1.24101	0.89	11,111.24	10,410	702
3	4	5	4.42%	2.60%	1.82%		0.94	0.87	20.00%	1.52633	1.13	16,080.18	15,587	493
									Price of the Cap (Total of Caplets):			37,674	34,795	2,879

Fig. 9 Calculation of the settlement amount, the intrinsic value, and the option price

Literature and References to the External Tool

Ernst, D., Häcker, J. (2021): Risk management in business step by step, UTB, Course 3—Assignment 3.
Bösch, M. (2020): Derivate: Verstehen, anwenden und bewerten, 4th edition, Munich, pp. 251-268.
Hull, J. C. (2014): Risk management: banking, insurance and other financial institutions, 3rd edition, Pearson, pp. 283-295.
See Excel file `Case Study Risk Management Part 2`, Excel worksheet `Hedging with Caps`

Assignment 21: Simulation-Based Planning: Definition of the Risk Parameters for the Monte Carlo Simulation of a Business Plan

Content

A company's risk management has the task of continuously managing risks through identification, quantification, aggregation, and monitoring. The law on control and Transparency Act (KonTraG) required the creation of transparency about the risks that the company are exposed to. The principles include:

- Crises that threaten the existence of the company should be avoided.
- In important business decisions the expected chances and risks are weighed against each other (Business Judgment Rule).

Risks are identified as possible deviations from the plan as part of corporate planning. This understanding of risk includes both:

- Opportunities (possible positive deviations) as well
- Dangers (possible negative deviations)

The principles of proper planning (GoP) support the close connection between planning and risk accounting by creating transparency of the risks that deviate from the plan. Analogous to IDW PS 340 n.F. for testing the early risk-detection system, which is based on the KonTraG, the GoP demand the provision of an aggregated total risk scope.

Deviations from plan (uncertainty in planning) and the associated threat to the existence of the company can be assessed. A corporate plan designed in

(continued)

compliance with these requirements of the GoP is thus the adequate basis for combining risk management and planning.

Liability risks for management, the board of directors or the supervisory board are avoided. The consideration of opportunities and dangers (risks) makes it possible to make meaningful decisions at all levels and to obtain decision-relevant plan values in the sense of expected values.

It is an essential requirement of the GoP to determine plan values that are "on average" correct, for which possible positive and negative plan deviations are adequately taken into account. These plan values are called unbiased plan values. Only planning values that are true to expectations can form the basis for business decisions (e.g., investment decisions or decisions about company acquisitions).

What is meant by unbiased planning and unbiased plan values?

The term "unbiasedness," in mathematical statistics, denotes a property of an estimating function. An estimator is said to be unbiased if its expectation is equal to the true value of the parameter to be estimated. The extent of the deviation of the expectation from the true value is called distortion or bias. The distortion is the systematic error of the estimator.

In corporate planning, unbiasedness means that, with the help of suitable planning and forecasting, methods are used to make the best possible, undistorted predictions in many planning cases and planning periods. A risk analysis is then a necessary prerequisite for deriving unbiased values.

Planned values without information about the planning uncertainty (e.g., standard deviation of plan deviations) are meaningless, as theoretical plan deviations can be of any size with relevant probability. In practice, however, it has been shown that plan values are often purely pseudo-deterministic plan values. With these, an attempt is made to reduce the complexity of the uncertainty to get "under control" and to specify values from which one accepts that these "most probable values" will occur.

We see that there is a significant difference between

- Unexpected plan values that reflect the expected development of a company map.
- The pseudo-deterministic plan values (often also referred to as plan values), which serve as roughly simplified planning aids.

If companies work with target values as control parameters, they should be derived and determined on the basis of forecast plan values. Only so can opportunities and dangers (risks) be identified, which can lead to deviations from plan in the future.

In the following, three assignments you will learn how unbiased and simulation-based planning are created. We proceed in the following steps:

(continued)

- Determination of the risk parameters for the Monte Carlo simulation of a business plan.
- Adoption of the risk parameters for the Monte Carlo simulation in the business plan.
- Risk aggregation using Monte Carlo simulation and risk analysis.

Assignment

- Open the Excel file `Case Study Risk Management Part (3)`.
- Create a new worksheet `Monte Carlo Parameters` and insert them after the worksheet `Assumptions`.
- Use the following parameters as the risk input variables:
- Sales growth: The sales growth has a triangular distribution with the following input values:
  - year t(1): minimum = 1.0%, most likely value = 3.0%, maximum = 3.5%
  - year t(2): minimum = 1.0%, most likely value = 3.0%, maximum = 3.5%
  - year t(3): minimum = 1.0%, most likely value = 3.0%, maximum = 3.5%
  - year t(4): minimum = 0.5%, most likely value = 2.5%, maximum = 3.0%
  - year t(5): minimum = 0.5%, most likely value = 2.0%, maximum = 2.5%
- Cost of goods sold (COGS): The cost of goods sold has a normal distribution with an expected value of 0% and a standard deviation of 5% for years t(1) to t(5).
- Selling expenses: The selling expenses show a normal distribution with an expected value of 0% and a standard deviation of 3% for the years t(1) to t(5).
- General administrative expenses: The general administrative expenses have a normal distribution with an expected value of 0% and a standard deviation of 2%.
- Other operating expenses: The other operating expenses are dependent on the sales, but can be affected by the defaults on receivables or losses of key accounts.
- Default of receivables: A distinction is made between small defaults, medium defaults, and large defaults.
- All defaults have a uniform distribution with the following parameters:
  - Small defaults: Have a 30% likelihood of damage and lead to a loss of 0.5% of the sales.
  - Medium defaults: Have a 5% likelihood of damage and lead to a loss amount of 1% of the sales.
  - Major defaults: Have a 1% likelihood of damage and lead to a loss of 3% of the sales.
- Loss of key accounts: Also has a uniform distribution, has a 5% likelihood of damage and leads to a loss of 2% of the turnover.

Content

The quantification of risks and their aggregation in the context of corporate planning is a particularly important business management task. This is where the corporate finance division and the risk management division work together. To describe a risk quantitatively, a probability distribution can be used, which describes the effects of a risk on earnings in a period (e.g., year).

The most important distribution functions in risk management are the binomial distribution, the normal distribution and the triangular distribution. These distributions describe the monetary effects of the risk in 1 year and, thus, integrate the frequency of the occurrence and the extent of the effects of the risk.

The *binomial distribution* is often used in practice. It describes a risk only by the amount of damage and the probability of occurrence. This is appropriate when "event-oriented risks" are considered with a certain effect. With these, one can approximately assume that the corresponding risk occurs exactly once in a year with the probability p and then results in damage. Typical applications are the loss of a key account, a fire in a factory or the failure of a critical machine. Event-oriented risks are thus either an "opportunity" or a "danger"—but not both at the same time. If an event can occur more than once within a year, on the other hand, a Poisson distribution or a general binomial distribution is required ($n > 1$).

Risks that represent the opportunity and the danger at the same time can be described, for example, by the *normal distribution*. For their specification, one needs the expected value, which as a location parameter indicating what happens "on average," and the standard deviation, which specifies the extent of "usual" positive or negative deviations. The normal distribution is used, in particular, to describe the risks that can be understood as an amalgamation of many individual small (and independent) individual events, such as the fluctuations in demand, sales fluctuations, interest rate, and currency risks and the changes in raw material prices (i.e., especially "market-related" risks).

In the simplest case, the so-called *triangular distribution* can be used to describe the asymmetrical risks that have either an opportunity or a danger overhang. In this case, a risk-related variable (e.g., the costs of a project) considered is described by (a) a minimum value, (b) a most probable value, and (c) a maximum value. Examples: risk-related possible bandwidth of the investment amount.

This small overview of the relevant probability distributions is, of course, not conclusive. The literature provides explanations of many other probability distributions that are important in practice, such as the lognormal distribution, the exponential distribution, the Poisson distribution (for describing the frequency of a risk occurrence), the uniform distribution (for situations where no

(continued)

information on the occurrence probabilities is available) and the Pareto distribution, which is suitable for modeling extreme events (such as the natural disasters or stock market crashes).

Important Formulas

No special knowledge of any formula is required for the preparatory measures. The inputs in the Excel and the selected Monte Carlo simulation software are the most relevant.

Procedure

Preliminary note: We will carry out the Monte Carlo simulation here using the Risk Kit software program, as an example. Any other, suitable software can, of course, also be used. Again, it should be noted that the results of the Monte Carlo simulation are only displayed if the corresponding software is installed on the computer.

- You have a business plan from PHARMA GROUP in Excel. This consists of the following worksheets:
  - Assumptions
  - Income statement
  - Assets
  - Equity and liabilities

The first step is to create the assumptions for the Monte Carlo simulation in a new worksheet and name it `Monte Carlo Parameter`.

- The first group of risk parameters relates to the sales and the related costs (cells `B7:J21`).
- The first relevant risk parameter is the sales growth (cells `B8:J11`).
- The sales growth has a triangular distribution with the following input values:
  - year t(1): minimum = 1.0%, most likely value = 3.0%, maximum = 3.5%
  - year t(2): minimum = 1.0%, most likely value = 3.0%, maximum = 3.5%
  - year t(3): minimum = 1.0%, most likely value = 3.0%, maximum = 3.5%
  - year t(4): minimum = 0.5%, most likely value = 2.5%, maximum = 3.0%
  - year t(5): minimum = 0.5%, most likely value = 2.0%, maximum = 2.5%
- In Risk-Kit, under `One-dimensional Distributions`, select ➲ `Triangular` ➲ `Single Random Number` from the triangular distribution and enter for the triangular distribution enter the following values per year for the input values for sales growth:
  - A: minimum growth rate
  - B: most probable growth rate
  - C: maximum growth rate of sales

- Mark the sales growth for all years (cells F8:J8) as input cells for the Monte Carlo simulation. To do this, go to Input in Risk-Kit. The input cells receive a green background color.
- The second relevant risk parameter is the costs of goods sold (COGS) (cells `B13:J15`). The costs of goods sold show a normal distribution for years t(1) to t (5) with an expected value of 0% and a standard deviation of 5%.
- The third relevant risk parameter is the selling expenses (cells `B16:J18`). The selling expenses show a normal distribution for years t(1) to t(5) with an expected value of 0% and a standard deviation of 3%.
- The fourth relevant risk parameter is the general administrative expenses (cells `B19:J21`). The general administrative expenses show a normal distribution with an expected value of 0% and a standard deviation of 2%.
- The second group of risk parameters concerns the other operating expenses (cells `B23:J47`). The other operating expenses are dependent on the sales, but may also be affected by the defaults on receivables or losses of key accounts.
- The first risk parameter in this group concerns the default of receivables (cells `B24:J41`). A distinction is made between the small defaults, the medium defaults, and the large defaults.
- All the damages have a uniform distribution with the following parameters.
  - Small default: Has a 30% likelihood of damage. If the random number drawn in the Monte Carlo simulation is less than or equal to 30%, this results in a loss of 0.5% of the sales `F29=IF(F24<=F27;F28;0)`.
  - Medium default: Has a 5% likelihood of loss. If the random number drawn in the Monte Carlo simulation is less than or equal to 5%, this results in a loss amount of 1% of the sales. `F35=IF(F30<=F33;F34;0)`.
  - Large failure: Has a 1% likelihood of damage. If the random number drawn in the Monte Carlo simulation is less than or equal to 1%, this results in a loss amount of 3% of the sales `F41=IF(F36<=F39;F40;0)`.
- The second risk parameter in this group concerns the loss of a key account (cells `B42:J47`). It also has a uniform distribution. If the random number drawn in the Monte Carlo simulation is less than or equal to 5%, this leads to a loss of 2% of the sales `F47=IF(F42<=F45;F46;0)`.

Result (Fig. 10)

Risk Assumptions	Plan t_1	Plan t_2	Plan t_3	Plan t_4	Plan t_5
Triangular distribution	Distribution	Distribution	Distribution	Distribution	Distribution
Sales growth	0.0328652	0.031584563	0.021499734	0.023045766	0.018501922
- Minimum value	1.00%	1.00%	1.00%	0.50%	0.50%
- Most likely value	3.00%	3.00%	3.00%	2.50%	2.00%
- Maximum value	3.50%	3.50%	3.50%	3.00%	2.50%
Normal distribution	Distribution	Distribution	Distribution	Distribution	Distribution
COGS	-0.062434985	-0.037692881	0.005279785	0.078803974	-0.019195299
- Expected value	0.00%	0.00%	0.00%	0.00%	0.00%
- Standard deviation	5.00%	5.00%	5.00%	5.00%	5.00%
Selling expenses	-1.75%	-1.19%	1.57%	-0.66%	-1.38%
- Expected value	0.00%	0.00%	0.00%	0.00%	0.00%
- Standard deviation	3.00%	3.00%	3.00%	3.00%	3.00%
General administration expenses	0.90%	-0.29%	-1.06%	-1.50%	1.48%
- Expected value	0.00%	0.00%	0.00%	0.00%	0.00%
- Standard deviation	2.00%	2.00%	2.00%	2.00%	2.00%
Uniform distribution	Distribution	Distribution	Distribution	Distribution	Distribution
Other operating expenses					
- Small defaults	87.81%	89.89%	88.42%	13.75%	84.68%
Probability of occurence (minimum)	0.00%	0.00%	0.00%	0.00%	0.00%
Probability of occurence (maximum)	100.00%	100.00%	100.00%	100.00%	100.00%
Likelihood of damage	30.00%	30.00%	30.00%	30.00%	30.00%
Amount of damages in % of sales	0.50%	0.50%	0.50%	0.50%	0.50%
Expected amount of damages in % of sales	0.00%	0.00%	0.00%	0.50%	0.00%
- Medium defaults	82.57%	32.68%	47.53%	71.57%	28.21%
Probability of occurence (minimum)	0.00%	0.00%	0.00%	0.00%	0.00%
Probability of occurence (maximum)	100.00%	100.00%	100.00%	100.00%	100.00%
Likelihood of damage	5.00%	5.00%	5.00%	5.00%	5.00%
Amount of damages in % of sales	1.00%	1.00%	1.00%	1.00%	1.00%
Expected amount of damages in % of sales	0.00%	0.00%	0.00%	0.00%	0.00%
- Large defaults	50.39%	62.53%	57.25%	9.60%	45.08%
Probability of occurence (minimum)	0.00%	0.00%	0.00%	0.00%	0.00%
Probability of occurence (maximum)	100.00%	100.00%	100.00%	100.00%	100.00%
Likelihood of damage	1.00%	1.00%	1.00%	1.00%	1.00%
Amount of damages in % of sales	3.00%	3.00%	3.00%	3.00%	3.00%
Expected amount of damages in % of sales	0.00%	0.00%	0.00%	0.00%	0.00%
- Loss of key accounts	4.30%	80.35%	5.69%	35.26%	54.53%
Probability of occurence (minimum)	0.00%	0.00%	0.00%	0.00%	0.00%
Probability of occurence (maximum)	100.00%	100.00%	100.00%	100.00%	100.00%
Likelihood of damage	5.00%	5.00%	5.00%	5.00%	5.00%
Amount of damages in % of sales	2.00%	2.00%	2.00%	2.00%	2.00%
Expected amount of damages in % of sales	2.00%	0.00%	0.00%	0.00%	0.00%

Fig. 10 Monte Carlo risk parameters

Literature and References to the Excel Tool

Ernst, D. (2020): Corporate Risk Management, Course 2—Assignment 18.

Standard literature:

Hopkin, P. (2018): Fundamentals of Risk Management: Understanding, Evaluating and Implementing Effective Risk, 5. Edition, New York.

Hull, J.C. (2018): Risk Management and Financial Institutions, 5. Edition, Hoboken (New Jersey).

Siehe Excel-Datei Case Study `Risk Management Part 2`

Assignment 22: Generation of Distribution Functions by Expert Interviews

Task

- Based on the information provided by four experts, create an aggregate probability distribution for the company's revenue growth.
- Open the Excel file Case Study Risk Management Part 3.
- Use the following parameters as risk input variables:
- *Expert 1:* Sales growth has a triangular distribution with the following input values:
  - Year t(1): Minimum = 1.25%, Prob.Value = 3.0%, Maximum = 3.5%
  - Year t(2): Minimum = 0.75%, Prob.Value = 2.5%, Maximum = 3.0%
  - Year t(3): Minimum = 0.50%, Prob.Value = 2.0%, Maximum = 2.5%
  - Year t(4): Minimum = 0.25%, Prob.Value = 1.5%, Maximum = 2.0%
  - Year t(5): Minimum = 0.25%, Prob.Value = 1.0%, Maximum = 1.5%
- *Expert 2:* Sales growth has a triangular distribution with the following input values:
  - Year t(1): Minimum = 2.00%, Prob.Value = 2.5%, Maximum = 2.75%
  - Year t(2): Minimum = 1.50%, Prob.Value = 2.0%, Maximum = 2.25%
  - Year t(3): Minimum = 0.50%, Prob.Value = 1.5%, Maximum = 1.75%
  - Year t(4): Minimum = 0.50%, Prob.Value = 1.0%, Maximum = 1.25%
  - Year t(5): Minimum = 0.25%, Prob.Value = 1.0%, Maximum = 1.25%
- *Expert 3:* Sales growth has a PERT distribution with the following input values:
  - Year t(1): Minimum = 2.0%, Prob.Value = 2.75%, Maximum = 3.0%
  - Year t(2): Minimum = 1.5%, Prob.Value = 2.25%, Maximum = 2.5%
  - Year t(3): Minimum = 0.5%, Prob.Value = 1.75%, Maximum = 2.0%
  - Year t(4): Minimum = 0.5%, Prob.Value = 1.25%, Maximum = 1.5%
  - Year t(5): Minimum = 0.5%, Prob.Value = 1.25%, Maximum = 1.5%
- *Expert 4:* Sales growth has a normal distribution with the following input values:
  - Year t(1): Expected value = 2.5%, Standard deviation = 1.5%
  - Year t(2): Expected value = 2.0%, Standard deviation = 1.5%
  - Year t(3): Expected value = 1.5%, Standard deviation = 1.5%
  - Year t(4): Expected value = 1.0%, Standard deviation = 1.5%
  - Year t(5): Expected value = 1.0%, Standard deviation = 1.5%
- The following weights are assigned to the experts' assessments:
  - *Expert 1:* 45%
  - *Expert 2:* 25%
  - *Expert 3:* 20%
  - *Expert 4:* 10%

- Calculate the aggregate sales growth as a weighted average of expert estimates.
- Perform a Monte Carlo simulation and calculate the mean value from the simulation results.
- Calibrate the Monte Carlo simulation results and derive a continuous distribution function.

Content

Crucial to the success of a Monte Carlo simulation are meaningful Monte Carlo parameters. We need to avoid the "garbage in, garbarge out" situation, as the Monte Carlo simulation should help make better decisions. This requires a valid initial data on the risk parameters. In reality, companies often lack an empirical data basis. The reasons for this are:

- the company is new (e.g. start-up company) or has a business model that is difficult to compare (company in a niche segment),
- the internal data required for the simulation was not collected by the company in the past,
- the available data is insufficient or only of a qualitative nature,
- the data collected in the past is no longer meaningful and relevant for forecasting in the future (e.g. due to changes in the business model, technical changes, changes in demand, etc.),
- the costs of obtaining the data are not commensurate with the benefits,
- the competencies required for the data acquisition and the evaluation are not available in the company.

The only way to generate meaningful data in the sense of the distribution functions for risk parameters is through expert surveys. The expert surveys use structured interviews with experts to determine the knowledge that cannot be generated in any other way. These expert surveys are based on the personal experience and the intuition of the questioned experts. However, this does not mean that this information is any less valid than historical data. Often the opposite is the case, as the experts can assess the company and its development better than it can be expressed by historical data. Furthermore, it is also possible to combine the past-oriented, empirical data samples with the future-oriented expert opinions.

Within the framework of the expert interviews, it is recommended that only the easily interpretable probability and the density functions be used to determine the risk ranges. These should be flexible and easy to adapt by the respective expert. This eliminates those distribution types, whose parameters have no direct connection to the distribution shape.

The probability distributions produced by the experts are aggregated by a Monte Carlo simulation and combined into an overall probability distribution.

(continued)

Thus, the different expert estimates can be used in aggregate. In this process, the possible expert estimates are randomly selected and combined. Subsequently, an expression of the random variable, i.e. a possible growth in sales, is calculated using the distribution selected in this way. The uncertain variable turnover can now be described by a given distribution, whose parameters are themselves uncertain (second order probability distribution). The probability distribution determined in this way captures the uncertainty of each expert's estimate of the sales growth, but also the divergence of different experts' estimates. Finally, a continuous distribution of sales growth can be determined by a calibration from the histogram of the sales growth.

Important Formulas

No special knowledge of any formula is required for the preparatory measures. The inputs in the Excel and the selected Monte Carlo simulation software are the most relevant.

Procedure

Preliminary remark: Here, we carry out the Monte Carlo simulation as an example with the software program Risk-Kit. Any other suitable software can, of course, also be used. It should be noted here again that the results of the Monte Carlo simulation are only displayed if the corresponding software is also installed on the computer.

- Using the data given in the assignment, create a triangular distribution (cells `F59:J59` and cells `F64:J64`) for Expert 1 and Expert 2, a PERT distribution (cells `F69:J69`) for Expert 3, and a normal distribution (cells `F74:J74`) for Expert 4.
- Mark the expert estimates for sales growth for all years (cells `F59:J59`, `F64:J64`, `F69:J69`, `F74:J74`) as input cells for the Monte Carlo simulation. To do this, go to `Input` in Risk-Kit. The input cells are given a green background color.
- Insert the weights for the experts' statements (cells `E59`, `E64`, `E69`, `E74`).
- Calculate the aggregate revenue growth from the expert opinions `F78=$E$59*F59+$E$64*F64+$E$69*F69+$E$74*F74)`. The output cells for the Monte Carlo simulation are activated in Risk Kit via `Output` and have a dark yellow background color in Risk Kit.
- In cells `F78:J78`, the mean value of the sales growth from the simulation is calculated. This assumed that the value is an unbiased value, as it takes into account all scenarios from the Monte Carlo simulation and thus all possible positive and negative deviations from the mean value. This assumption value, which is true to the expectations, can therefore be regarded as correct "on average."
- Start the Monte Carlo simulation by pressing the icon 

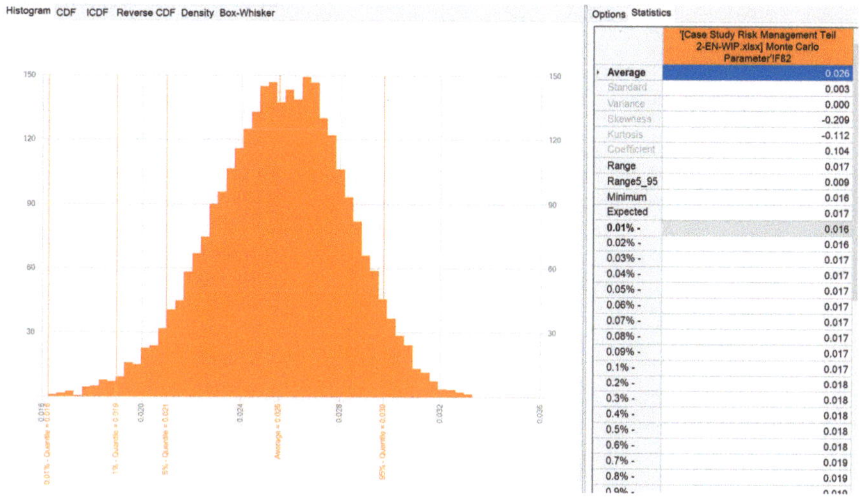

Fig. 11 Histogram of the simulation results

- IMPORTANT: After running the Monte Carlo simulation, the calculation options automatically reset the workbook calculation to Manual. Please set this back to `Automatic`.
- Cells `F89:J5088` list the results of the Monte Carlo simulation. The values originate from the `Output` worksheet and are copied from it.
- Figure 11 shows an example of the frequency distribution of the simulation results for year t(1).

- Through calibration, a continuous distribution function can also be derived from the simulation results.

- To do this, we press the icon 

and select Calibrate one-dimensional distributions.
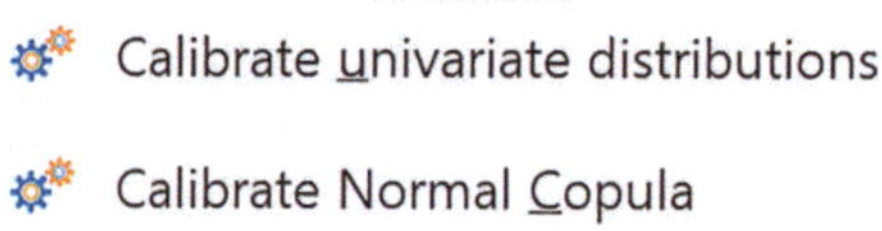

- A window then opens in which we make the following settings. After that the calibration is carried out (Fig. 12).

- The following window (Fig. 13) now appears, in which you press `OK`.

- The result of the calibration shows that the Weibull distribution best reflects the distribution of aggregate sales growth rates (Fig. 14).

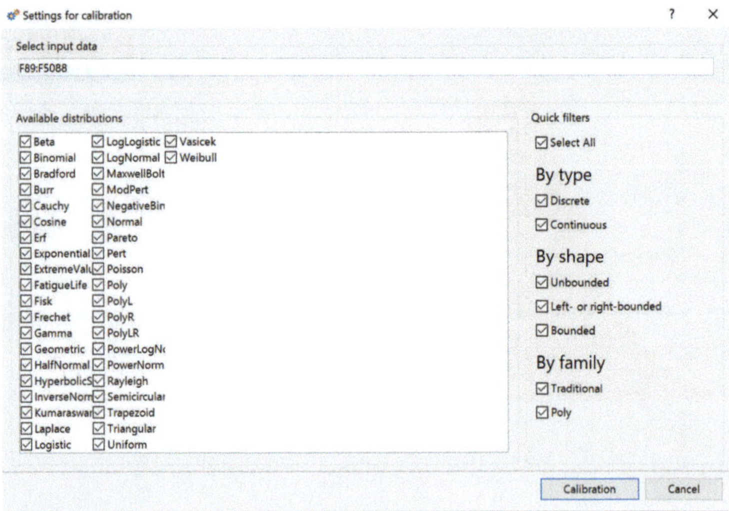

Fig. 12 Calibration

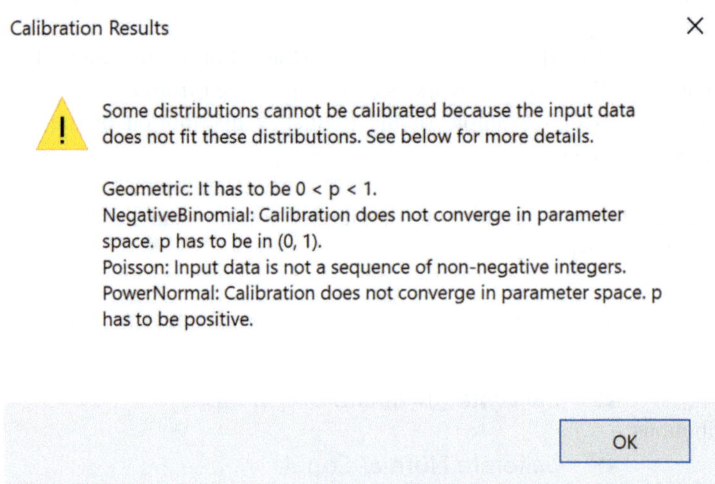

Fig. 13 Note on the calibration result

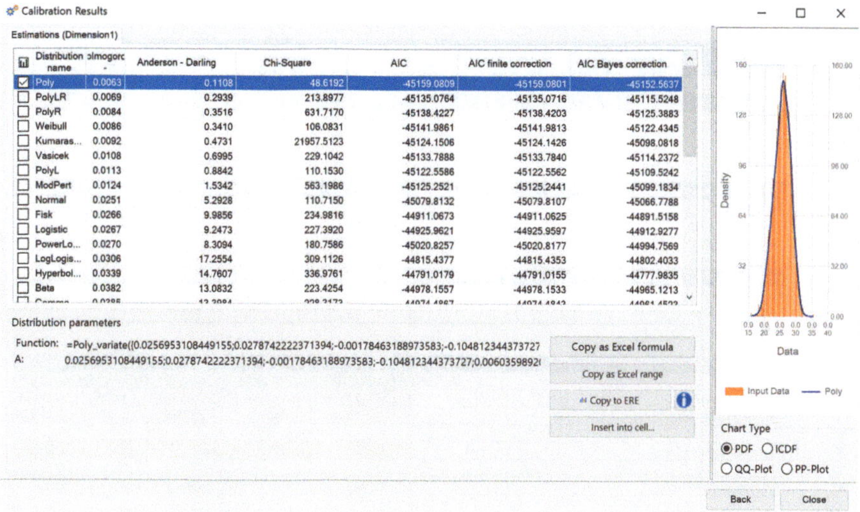

Fig. 14 Calibration result

- Now select the field Insert into a cell... and then enter the cell F82.

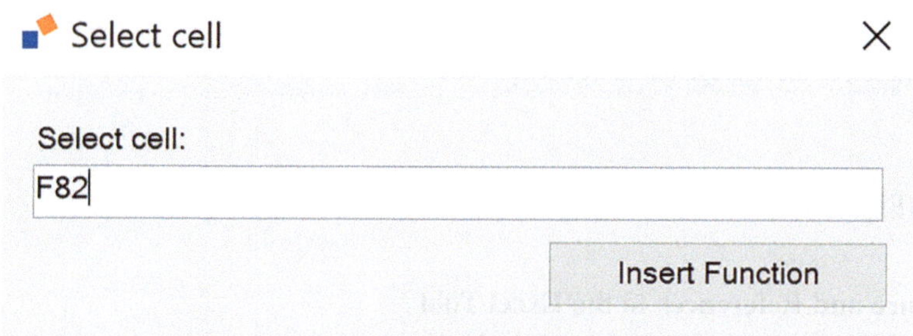

- Cells `F82:J82` now contain aggregated estimates of the experts for the sales growth. These can, now, be used for the following assignments. For the reasons of simplification, however, we will continue to work with the triangular distribution in the cells `F8:J8`. As mentioned, you can also use the cells `F82:J82`.

		Plan t_1	Plan t_2	Plan t_3	Plan t_4	Plan t_5
Expert 1: Triangular distribution	Weight	Distribution	Distribution	Distribution	Distribution	Distribution
Sales growth	45%	1.87%	1.84%	2.12%	0.95%	0.88%
- Minimum value		1.25%	0.75%	0.50%	0.25%	0.25%
- Most likely value		3.00%	2.50%	2.00%	1.50%	1.00%
- Maximum value		3.50%	3.00%	2.50%	2.00%	1.50%
Experte 2: Triangular distribution	Weight	Distribution	Distribution	Distribution	Distribution	Distribution
Sales growth	25%	2.28%	2.10%	0.58%	0.82%	0.42%
- Minimum value		2.00%	1.50%	0.50%	0.50%	0.25%
- Most likely value		2.50%	2.00%	1.50%	1.00%	1.00%
- Maximum value		2.75%	2.25%	1.75%	1.25%	1.25%
Expert 3: PERT distribution	Weight	Distribution	Distribution	Distribution	Distribution	Distribution
Sales growth	20%	2.73%	2.07%	1.73%	1.07%	1.06%
- Minimum value		2.00%	1.50%	0.50%	0.50%	0.50%
- Most likely value		2.75%	2.25%	1.75%	1.25%	1.25%
- Maximum value		3.00%	2.50%	2.00%	1.50%	1.50%
Expert 4: Normal distribution	Weight	Distribution	Distribution	Distribution	Distribution	Distribution
Sales growth	10%	2.54%	3.15%	4.12%	1.69%	3.27%
- Expected value		2.50%	2.00%	1.50%	1.00%	1.00%
- Standard deviation		1.50%	1.50%	1.50%	1.50%	1.50%
Aggregate sales growth		2.21%	2.08%	1.86%	1.02%	1.04%
Revenue growth (mean)		2.55%	2.05%	1.53%	1.13%	0.95%
Calibrated distribution function		2.62%	2.32%	2.71%	2.47%	1.93%

Simulation results

	Plan t_1	Plan t_2	Plan t_3	Plan t_4	Plan t_5
	2.37%	2.01%	1.51%	0.89%	0.79%
	2.61%	1.95%	1.26%	0.85%	0.55%
	2.55%	2.14%	1.71%	0.76%	1.03%
	2.31%	1.80%	0.65%	1.46%	1.12%

Fig. 15 Result of the aggregation of expert opinions

Result (Fig. 15)

Literature and References to the Excel Tool

Gleißner, W. (2008): Erwartungstreue Planung und Planungssicherheit - Mit einem Anwendungsbeispiel zur risikoorientierten Budgetierung, in: Controlling, Heft 02/2008, S. 81-87.

Klein, Martin (2010): Monte-Carlo Simulation and Due Diligence: A Methodological Approach to Computer-Aided Aggregation of Probability Distributions from Expert Surveys, Working Papers in Accounting Valuation Auditing 2010-5, Friedrich-Alexander University Erlangen-Nuremberg, Chair of Accounting and Auditing.

See Excel file `Case Study Risk Management Part 3`, Excel worksheet `Monte Carlo Parameters`

Assignment 23: Transfer of the Risk Parameters for the Monte Carlo Simulation into the Business Plan

Assignment

The following are defined as the risk input variables:

- Total sales
- Costs of goods sold (COGS)
- Selling expenses
- General administrative expenses
- Other operating expenses

Insert rows with the following names for each risk input variable:

- Random value: The random value is the value according to a combination of stochastic Monte Carlo parameter.
- Expected value: The expected value is the average over all scenarios created by Monte Carlo simulation.
- Planned value: The planned value is the value according to the business plan.
- Risk value: The risk value is the deviation of the planned value from the expected value.

Linking the Monte Carlo parameters with the planning calculation:

- Connect the parameters of the Monte Carlo simulation with the income statement planning.
- For each risk input variable, the planned value, the expected value, and the risk are calculated. The planned value is the value according to the business plan, the expected value is the risk-based value and the risk is the deviation of the planned value from the expected value.
- Adjust the entire plan (income statement, balance sheet, and cash flow statement) so that the expected values are used instead of the planned values.

Content

The probability distributions that we have defined for the relevant, unhedged risk input variables must now be integrated into the business plan. For this purpose, the probability distributions are linked to the items of the income statement planning.

The following values are calculated for each risk input variable:

- Random value: The random value is the value according to a combination of stochastic Monte Carlo parameter.
- Planned value: The planned value is the value according to the business plan.

(continued)

- Expected value: The expected value is the risk-based value.
- Risk value: The risk value is the deviation of the planned value from the expected value.

The future orientation requires planning and this is associated with the risks. The risk analysis, as the most important element of risk management, identifies and quantifies risks. They are the necessary prerequisite for the derivation of "unbiased plan values" and the assessment of planning (un)-certainty as the extent of the possible plan deviations. The expected plan values occur "on average" overall risk-related possible future scenarios and their calculation requires a risk analysis.

We note that the (unbiased) planned values, which describe the expected development of a company, differ significantly from the target values (often also referred to as planned values) for corporate management and possibly also for remuneration systems. Only "unbiased plan values" are a basis for entrepreneurial decisions. The requirement for unbiased plan values (estimates) means that the best possible undistorted forecasts are made with the aid of suitable planning and forecasting systems and that these provide information about the possible future values and the associated probabilities. In practice, however, it is apparent that plan values are often specified purely as target values. These target values are in no way suitable for risk-based planning. This is not only due to the methodological deficits, but also to the fact that the target values are associated with a desired message and thus a pseudo-deterministic scenario is to be created. For example, "conservative planning" is deliberately created so that planned values are most likely to be reached or exceeded. Conversely, "demanding" or "challenging" planning is also found at least as frequently, in which the ambitious plan values are set as targets that can only be achieved with the greatest effort and favorable framework conditions. Combinations of both the methods are also very often found at different levels or in different areas of the same company due to the lack of uniform and/or clear guidelines in the corporate planning or controlling. All the planning approaches mentioned above do not allow any forecasts about the actual development of the company to be made. Therefore, they are not a suitable basis for the business decisions (e.g., in the capital budgeting).

In order to generate unbiased plan values, we use random variables based on the probability distributions as the input variables for unhedged risks and aggregate them using the Monte Carlo simulation to generate unbiased output variables. We first identify the unhedged risks, define distribution functions, and then incorporate the risk-based variables into the business plan.

Important Formulas

No special knowledge of any formula is required for the calculation of the planned values, the expected values, and the risk values.

Procedure

Insert for the following rows:

- Total sales (random value) (Row 7)
- Total sales (expected value) (Row 8)
- Total sales (planned value) (Row 9)
- Total risk (risk value) (Row 10)

- Cost of goods sold (random value) (Row 12)
- Costs of goods sold (COGS) (expected value) (Row 13)
- Costs of goods sold (COGS) (planned value) (Row 14)
- Costs of goods sold (COGS) (risk value) (Row 15)

- Gross profit on sales (random value) (Row 17)
- Gross profit on sales (expected) (Row 18)
- Gross profit on sales (planned value) (Row 19)

- Gross profit on sales (risk value) (Row 20)
- Selling expenses (random value) (Row 22)
- Selling expenses (expected value) (Row 23)
- Selling expenses (planned value) (Row 24)
- Selling expenses (risk value) (Row 25)

- General administrative expenses (random value) (Row 29)
- General administrative expenses (expected value) (Row 30)
- General administrative expenses (planned value) (Row 31)
- General administrative expenses (Row 32)

- Other operating expenses (random value) (Row 36)
- Other operating expenses (expected value) (Row 37)
- Other operating expenses (planned value) (Row 38)
- Other operating expenses (risk value) (Row 39)

- EBIT (random value) (Row 42)
- EBIT (expected value) (Row 43)
- EBIT (planned value) (Row 44)

- EBIT (risk value) (Row 45)
- EBT (expected value) (Row 51)

- EAT (expected value) (Row 55)

- Net income (expected value) (Row 59)

- Sales (random) (cells `F7:J7`) are linked to the Monte Carlo parameters "sales growth" (cells `F8:J8`). The sales (random) in t(1) (cell `F7`) are linked to the sales from year t(0) `F7=E9*(1+'Monte Carlo Parameter'!F8)`. From year t (2) onward, the sales are linked to the sales (expected) from the previous year `G7=F7*(1+'Monte Carlo Parameter'!G8)`. As a result, the development of sales revenue is dependent on both the random variable "growth rate" and the random variable "sales revenue (random)." This results in a higher volatility compared to the situation where the sales (expected) depend on the planned sales. The sales revenues (expected values) are calculated as the mean value over the planned random values F8=Mean(F7). Row 9 contains the planned values from corporate planning, which are not based onstochastic planning. The deviations between the random values and the planned values result in the risk values, which can be found for sales in line 10. The breakdown into random value, expected values, planned values and risk values is also used for the following income statement items.

- The costs of goods sold (expected), the selling expenses (expected), and the general administrative expenses (expected) are each modelled in a similar way. For example, the costs of goods sold in t(1) (cell `F12`) are modeled by multiplying the sales (expected) with the cost of goods sold ratio (cost of goods sold (in % sales) and the random parameter from the Monte Carlo parameters `F12=F7*Assumptions!F126*(1+'Monte Carlo Parameter'!F13)`.
- The formula for the other operating expenses is more complex and includes the damages due to the small defaults, the medium defaults, and the large defaults, as well as damages due to the loss of a key account. For t(1) the formula is `F36=F7*Assumptions!F136+GuV!F7*'Monte Carlo Parameter'!F29+F7*'Monte Carlo Parameter'!F35+F7*'Monte Carlo Parameter'!F41+F7*'Monte Carlo Parameter'!F47`.
- The formula consists of the following components:
  - Other operating expenses excluding exceptional damages: `F7*Annahmen!F136`
  - Damages from small defaults: `F7*'Monte Carlo Parameter'!F29`
  - Damages from medium defaults: `F7*'Monte Carlo Parameter'!F35`
  - Damages from small defaults: `F7*'Monte Carlo Parameter'!F41`,
  - Damages due to loss of a key account: `F7*'Monte Carlo Parameter'!F47`

Income Statement

Absolute figures in € million	Actual t_{-2}	Actual t_{-1}	Actual t_0	Plan t_1	Plan t_2	Plan t_3	Plan t_4	Plan t_5
Total sales (random value)				40,396	41,014	41,925	42,950	43,769
Total sales (expected value)				40,573	41,585	42,625	43,480	44,205
Total sales (planned value)	34,943	35,015	39,586	40,774	41,997	43,257	44,338	45,225
Total sales (risk value)				-200	-412	-631	-858	-1,020
- COGS (random value)				14,725	14,902	15,059	16,529	17,499
- COGS (expected value)				14,753	15,134	15,497	15,819	16,081
- COGS (planned value)	11,756	11,382	17,010	14,831	15,276	15,734	16,127	16,450
- COGS (risk value)				-78	-142	-236	-309	-368
Gross profit on sales (random value)				25,671	26,113	26,867	26,421	26,270
Gross profit on sales (expected value)				25,821	26,451	27,128	27,662	28,123
Gross profit on sales (planned value)	23,187	23,633	22,576	25,943	26,721	27,523	28,211	28,775
Gross profit on sales (risk value)				-122	-270	-395	-549	-852
- Selling expenses (random value)				12,256	13,784	13,935	13,557	14,300
- Selling expenses (expected value)				12,959	13,286	13,629	13,895	14,131
- Selling expenses (planned value)	11,148	11,116	12,751	13,029	13,419	13,822	14,168	14,451
- Selling expenses (risk value)				-69	-133	-193	-273	-320
- Research and development expenses	4,405	4,504	5,246	5,263	5,421	5,583	5,723	5,837
- General administration expenses (random value)				2,491	2,397	2,464	2,606	2,648
- General administration expenses (expected value)				2,414	2,472	2,535	2,586	2,629
- General administration expenses (planned value)	1,804	2,026	2,728	2,425	2,497	2,572	2,637	2,689
- General administration expenses (risk value)				-11	-25	-37	-51	-60
+ Other operating income	787	864	5,057	2,378	2,449	2,523	2,586	2,637
- Other operating expenses (random value)				1,924	1,748	1,997	2,690	2,084
- Other operating expenses (expected value)				1,860	1,911	1,957	1,999	2,029
- Other operating expenses (planned value)	879	948	2,994	1,738	1,790	1,844	1,890	1,928
- Other operating expenses (risk value)				122	122	114	109	102
EBIT (random value)				6,115	5,212	5,410	4,432	4,037
EBIT (expected value)				5,703	5,809	5,945	6,045	6,134
EBIT (planned value)	5,738	5,903	3,914	5,867	6,043	6,224	6,380	6,507
EBIT (risk value)				-164	-233	-279	-334	-374
+ Result from associates accounted for using the equity method	-6	20	68	0	0	0	0	0
+ Financial income	149	289	910	122	122	126	153	185
- Financial expenses	1,108	1,635	2,574	2,017	2,127	2,126	2,171	2,235
Earnings before taxes (EBT) (random value)				4,220	3,207	3,410	2,414	1,987
- Taxes on income and earnings	1,017	1,329	607	1,493	1,521	1,557	1,583	1,606
Annual results - EAT (random value)				2,726	1,686	1,853	831	381
- of which attributable to non-controlling interest	13	-1	16	11	7	8	3	2
Net income (random value)				2,715	1,679	1,845	828	379
Dividend				1,629	1,007	1,107	497	228
Retained Earnings (random value)				1,086	672	738	331	152

Fig. 16 P&L with plan random values, unbiased plan values, target values, and risk values

Result (Fig. 16)

Literature and References to the Excel Tool

Ernst, D. (2020): Corporate Risk Management, Course 2—Assignment 19.

Standard literature:

Hopkin, P. (2018): Fundamentals of Risk Management: Understanding, Evaluating and Implementing Effective Risk, 5. Edition, New York.

Hull, J.C. (2018): Risk Management and Financial Institutions, 5. Edition, Hoboken (New Jersey).

Siehe Excel-Datei Case Study `Risk Management Part 2`

Assignment 24: Risk Aggregation Using Monte Carlo Simulation and Risk Analysis

Assignment

- Define the EBIT of years t(1) to t(5) as the relevant output variable for your Monte Carlo simulation.
- Run the Monte Carlo simulation with 5000 drawings.
- Process the output values of the Monte Carlo simulation for EBIT for the years t (1) to t(5) by transferring them to the worksheet Risk Analysis (1).
- Calculate the deviations resulting in the planning period from the difference between the simulated EBIT and the planned EBIT.
- Calculate the Value at Risk (Relative Value at Risk) for a confidence level of 95% and for a term of 1 year based on the EBIT values in the planning period.
- Calculate the conditional Value at Risk (expected shortfall) for a term of one year based on the EBIT values in the planning period.

Content

From the risk parameters identified and integrated into the business plan, it can only be deduced which stand-alone risks endanger the existence of a company. However, it is usually precisely the combined effects of several individual risks that trigger crises or even insolvency. In order to assess how large the total risk volume is (and thus the probability of insolvency as a result of the quantity of all risks), a so-called risk aggregation is necessary.

The objective of risk aggregation is to determine the overall risk position of a company and to determine the relative importance of the individual risks, taking into account interactions (correlations) between them. This requires the aggregation of all (material) risks across all risk types and also over time. Since only the quantified risks can be aggregated, the requirement to quantify all the risks is only consistent. By aggregating the quantified risks in the context of planning—opportunities and risks as the cause of possible deviations from the plan—it must be examined what effects these have on the future earnings.

In the integrated company-wide risk management, the risk aggregation procedures must be chosen that:

- Can capture the risks described by any probability distribution.
- Be able to take into account the non-additive (e.g., multiplicative) links between risks.
- Establish the context for the corporate planning and the future rating, since the risk management ultimately aims to show the planning reliability and the liquidity and equity capital requirements of a company consistently with the planning (and illiquidity usually occurs when covenants rupture

(continued)

due to the combined effect of several risks or a minimum rating of "B" is no longer achieved).

Only risk simulation (Monte Carlo simulation) meets the above requirements because the risks—unlike costs and revenues—cannot be added together. These simulation methods are the further development of the well-known scenario analysis techniques. A computer simulation is used in the risk aggregation to calculate and analyze a large representative number of risk-related possible future scenarios (planning scenarios). In this way, a realistic range of the future earnings and the liquidity development is shown, i.e., the planning reliability or extent of possible negative deviations from the plan. The probability that the covenants will be breached or that a necessary target rating will no longer be achieved in the future can be derived directly.

Important Formulas

As with the Monte Carlo simulations already carried out, the formulas required here, when using the Monte Carlo simulation, do not relate to the generation of the probability distribution of the target figure, but to the calculation of the Value at Risk or conditional Value at Risk. The result of the Monte Carlo simulation is in turn a discrete probability distribution for EBIT in the planning period. The planned EBIT value for the respective planning year is deducted from the simulation values. On these values, which represent the risk of a deviation from the planned value, the already known formulas for the Value at Risk and the conditional Value at Risk are applied with a discrete probability distribution.

Formally speaking, the Value at Risk is the absolute value of the negative quantile of a distribution at the confidence level p. The formula for calculating the Value at Risk as a return for a discrete probability distribution is:

$$\boxed{\mathrm{VaR}_\alpha(R) = \mathrm{RVaR}_\alpha(R) = |Q_\alpha(R)|} \qquad (19)$$

R = Risk = Deviation of the simulated values of EBIT from the planned value of EBIT

$\mathrm{VaR}_\alpha(R)$ = Value at Risk at confidence level p for the risk

$\mathrm{RVaR}_\alpha(\mathrm{R})$ = Relative Value at Risk at confidence level p for the risk

$Q_\alpha(R)$ = Quantile of distribution to confidence level p

The formula for the calculation of the conditional Value at Risk as a return with a discrete probability distribution is:

$$\boxed{\mathrm{CVaR}_\alpha(R) = |E(R|R < \mathrm{Q}_\alpha(R))|} \qquad (20)$$

CVaR_α = Conditional Value at Risk at confidence level p for discrete returns

Excel workbook: `Case Study Risk Management Part (2)` ➲ Worksheet: `Risk Analysis (2)`

Calculation of the relative Value at Risk as a return with a discrete probability distribution in the Excel:

Excel example: `C10=ABS(MIN(VLOOKUP(C8;'Risk Analysis (1)'!F5:G5004;2;0);0))`

Calculation of the conditional Value at Risk (expected shortfall) as a return with a discrete probability distribution in the Excel:

Excel example: `C13=ABS(AVERAGE('Risk Analysis (1)'!G5:G253))`

Procedure

Preliminary note: We will carry out the Monte Carlo simulation here using the Risk Kit software program as an example. Any other suitable software can, of course, also be used. Again, it should be noted that the results of the Monte Carlo simulation are

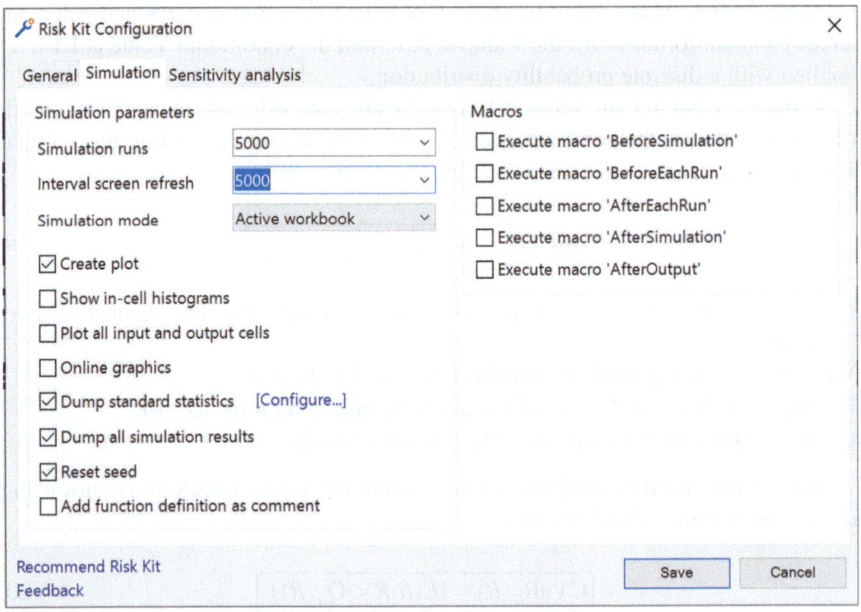

Fig. 17 Risk Kit configuration

only displayed if the corresponding software is also installed on the computer (Fig. 17).

- In the first step, the Monte Carlo simulation with 5000 drawings is carried out. Make sure that the results of the Monte Carlo simulation, that is, the 5000 output values of the EBIT, are displayed. These values are then copied into the worksheet `Risk Analysis (1)`. The EBIT figures for the year t(1) are in the cells `C5:C5004`.
- In the next step, the deviations from the planned value are determined `D5=C5-$D$2`. This deviation is the risk that the target value will not be achieved.
- The deviations from the planned value are only shown in an ascending order in the columns `F:G G5=SMALL($D$5:$D$5004;F5)`.
- Further risk analysis is done in the worksheet `Risk Analysis (2)`. On the basis of this list, the number of α-% smallest values is determined for a discrete probability distribution when determining the Value at Risk. In our example, these are the 5% smallest values. With 5000 available returns, this means 250 values. The 5% quantile is statistically defined as the 250th worst value. Our VaR approach uses the 250th value of the ordered list in column G of the worksheet `Risk Analysis (1)`. This is calculated with the function `C8=ROUNDDOWN((C6)*COUNT('Risk Analysis (1)'!G5:G5004);0)`. In our example, 5% of the smallest values are 250 values.
- The yield of the 250 smallest value is then determined. This is done for the 5% quantile with the function `C10=ROUNDDOWN((C6)*COUNT('Risk Analysis (1)'!F5:G5004);0)`. The calculated Value at Risk amounts to EUR 1664.
- Finally, the conditional Value at Risk can be calculated `C13=ABS(AVERAGE('Risk Analysis (1)'!G5:G253))`. It amounts to EUR 2051.

Result (Figs. 18 and 19)

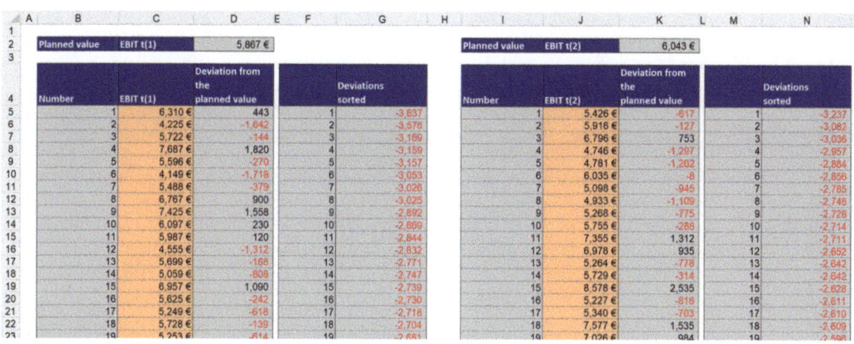

Planned value	EBIT t(1)	5,867 €		
Number	EBIT t(1)	Deviation from the planned value		Deviations sorted
1	6,310 €	443	1	-3,637
2	4,225 €	-1,642	2	-3,576
3	5,722 €	-144	3	-3,169
4	7,687 €	1,820	4	-3,159
5	5,596 €	-270	5	-3,157
6	4,149 €	-1,718	6	-3,053
7	5,488 €	-379	7	-3,026
8	6,767 €	900	8	-3,025
9	7,425 €	1,558	9	-2,892
10	6,097 €	230	10	-2,869
11	5,987 €	120	11	-2,844
12	4,555 €	-1,312	12	-2,832
13	5,699 €	-168	13	-2,771
14	5,059 €	-808	14	-2,747
15	6,957 €	1,090	15	-2,739
16	5,625 €	-242	16	-2,730
17	5,249 €	-618	17	-2,718
18	5,728 €	-139	18	-2,704

Planned value	EBIT t(2)	6,043 €		
Number	EBIT t(2)	Deviation from the planned value		Deviations sorted
1	5,426 €	-617	1	-3,237
2	5,916 €	-127	2	-3,082
3	6,796 €	753	3	-3,036
4	4,746 €	-1,297	4	-2,957
5	4,781 €	-1,262	5	-2,884
6	6,035 €	-8	6	-2,856
7	5,098 €	-945	7	-2,785
8	4,933 €	-1,109	8	-2,748
9	5,268 €	-775	9	-2,728
10	5,755 €	-288	10	-2,714
11	7,355 €	1,312	11	-2,711
12	6,978 €	935	12	-2,652
13	5,264 €	-778	13	-2,642
14	5,729 €	-314	14	-2,642
15	8,578 €	2,535	15	-2,628
16	5,227 €	-816	16	-2,611
17	5,340 €	-703	17	-2,610
18	7,577 €	1,535	18	-2,609

Fig. 18 Results of the Monte Carlo simulation

Year	t(1)	t(2)	t(3)	t(4)	t(5)
Confidence level	95%	95%	95%	95%	95%
Alpha	5%	5%	5%	5%	5%
Number of values	250	250	250	250	250
VaR = relativer VaR	1,664 €	1,683 €	1,744 €	1,700 €	1,848 €
Expected Shortfall	2,051 €	2,063 €	2,107 €	2,156 €	2,268 €

Fig. 19 Value at Risk and Expected shortfall

Literature and References to the Excel Tool

Ernst, D. (2020): Corporate Risk Management, Course 2—Assignment 20.

Standard literature:

Hopkin, P. (2018): Fundamentals of Risk Management: Understanding, Evaluating and Implementing Effective Risk, 5. Edition, New York.

Hull, J.C. (2018): Risk Management and Financial Institutions, 5. Edition, Hoboken (New Jersey).

Siehe Excel-Datei Case Study `Risk Management Part 2`

Index

D. Ernst, J. Häcker, *Corporate Risk Management*, Springer Texts in Business and Economics, https://doi.org/10.1007/978-3-031-53126-2

MIX
Papier aus verantwortungsvollen Quellen
Paper from responsible sources
FSC® C105338
www.fsc.org

If you have any concerns about our products,
you can contact us on
ProductSafety@springernature.com

In case Publisher is established outside the EU,
the EU authorized representative is:
Springer Nature Customer Service Center GmbH
Europaplatz 3, 69115 Heidelberg, Germany

Printed by Libri Plureos GmbH
in Hamburg, Germany